ENERGY
FOR WORLD
AGRICULTURE

FAO Agriculture Series No. 7

ENERGY FOR WORLD AGRICULTURE

by
B.A. Stout

in collaboration with

C.A. Myers
A. Hurand
L.W. Faidley

FOOD AND AGRICULTURE ORGANIZATION
OF THE UNITED NATIONS
Rome 1979

First printing 1979

Second printing (with corrections) 1981

P-05
ISBN 92-5-100465-X

FOREWORD

For some time the Food and Agriculture Organization of the United Nations has shown interest in the effective application of energy to agriculture. In 1961, the Organization produced its first publications on energy for developing countries: one on the utilization of solar energy and the other on windmills for water-lifting and electricity generation. More recently, an entire chapter of the 1977 FAO publication, *The state of food and agriculture*, was devoted to a discussion of energy in agriculture. The present publication demonstrates FAO's continuing efforts to provide guidance and information on this important subject.

Agricultural production is dependent upon photosynthesis for the conversion of solar energy into a form suitable for consumption by animals and man. In addition, modern agriculture requires fossil energy, in the form of fertilizer and pesticides, and fuel for the operation of agricultural machines. Agriculture is thus both a producer and a consumer of energy.

Two salient features of the world energy situation are the very modest share of developing countries in total consumption and the small proportion of energy used for agricultural production. The developing countries' share of world commercial energy is 18 percent; worldwide agriculture consumes approximately 3.5 percent, of which 2.9 percent is used by developed and 0.6 percent by developing countries.

Although agriculture accounts for only a small share of overall energy consumption, it clearly deserves to be given the highest priority in that regard, since the high-yield agriculture practised in developed countries and the increased yields expected from the application of "green revolution" technology in developing countries depend to a large extent on energy-intensive inputs such as fertilizer, machinery and pesticides. There is nonetheless a need for economy in the use of commercial energy in agricultural production, to avoid large increases in the price of food and other agricultural products. For the same reason, reduced energy consumption in the processing, transport, marketing and preparation of food must be achieved.

It seems, however, that the developing countries may have several options open to them other than the highly energy-intensive methods adopted by developed countries when commercial energy was inexpensive. As agriculture in the developing worlds is in the process of substantial change, there is still scope for the introduction of energy-efficient technologies.

The purpose of this publication is to put into perspective energy in relation to the food system and to provide guidance in the application of energy to agriculture for those concerned with increasing food supplies — especially in developing countries. Energy resources are set out in tabular form, and the principles and efficiencies of various conversion processes are considered together with detailed descriptions of their application to agriculture and the total food system.

A conceptual outline is given of the energy flow in the food system, and the energy requirements of each operation in food production, processing and delivery are set out. The implications, as far as energy is concerned, of providing an adequate diet for the expanding world population are discussed, as are alternative energy sources for the future.

One of the main advantages of this publication is its concise coverage of many energy-related subjects and its analyses of a wealth of data collected from all parts of the world.

I sincerely hope it will be of service to agriculturalists, engineers, planners and others concerned with making world agriculture efficient enough to meet the basic food needs of all nations.

EDOUARD SAOUMA

Director-General

ACKNOWLEDGEMENTS

Many individuals and agencies have contributed to this book. It is not possible here to acknowledge individually the hundreds of authors of the references cited, although their work, collectively, made this book possible. The sources of tables and figures and of text citations are indicated by reference numbers (see pages 259–275).

Grants from the Rockefeller Foundation, the Brookhaven National Laboratory, the Midwest Universities' Consortium for International Assistance and the U.S. Agency for International Development made possible the collection of hundreds of energy references and the preparation of a computerized information retrieval system.

The French Government supported A. Hurand while he worked on the project. Michigan State University granted B. Stout a sabbatical leave while the manuscript was being drafted. FAO deserves special thanks for recognizing the need for a document to put the energy and food situation in perspective, for assigning L. Faidley to work on the project and for providing support funds. H. von Hulst and W. van Gilst of FAO's Agricultural Engineering Services have been a constant source of assistance and encouragement throughout the project.

A number of reviewers of the draft manuscript offered valuable suggestions and constructive criticism. The authors gratefully acknowledge the following for their helpful comments.

W. Baader, Director
Institute of Agricultural Machinery Research
Federal Research Centre of Agriculture
Braunschweig, Federal Republic of Germany

John Balis
U.S. Agency for International Development
Washington, D.C.

Bill Chancellor
Agricultural Engineering Department
University of California
Davis, California

Larry Connor
Agricultural Economics Department
Michigan State University
East Lansing, Michigan

Peter Fraenkel
Intermediate Technology Development Group Ltd
London

Earle Gavett
Economic Research Service
U.S. Department of Agriculture
Washington, D.C.

Iwao Ishino
International Studies and Programs
Michigan State University
East Lansing, Michigan

Yoshikuni Kishida Jr
Shin-Norinsha Co. Ltd
Tokyo

T.A. Lawand, Director, Field Operations
Brace Research Institute
Ste. Anne de Bellevue
Quebec, Canada

Gerald Leach
International Institute for Environment and Development
London

A.M. Michael
Project Director
Water Technology Centre
Indian Agricultural Research Institute
New Delhi, India

Francis X. Murray
Center for Strategic and International Studies
Georgetown University
Washington, D.C.

W. Harold Parady
Executive Director
American Association for Vocational Instructional Materials
Athens, Georgia

David Pimentel
Department of Entomology
Cornell University
Ithaca, New York

G.J. Quast
Department of Agricultural Engineering
Agricultural University
Wageningen, the Netherlands

G. Segler
Institut für Agrartechnik
Universität Hohenheim
Stuttgart, Federal Republic of Germany

C.P. Singh
Coordinating Engineer
Department of Farm Machinery and Power
Punjab Agricultural University
Ludhiana, India

Frederick A. Smith
Shell Research Ltd
Sittingbourne, Kent, U.K.

Douglas Williams
Department of Soils, Water and Engineering
University of Arizona
Tucson, Arizona

Sylvan Wittwer, Director
Agricultural Experiment Station
Michigan State University
East Lansing, Michigan

Lastly, a project of this type would have been impossible without dedicated, competent research assistants, draftsmen, typists, editors and other helpers — with special thanks to James Anderson, Sue Cooley, Carolyn Cunningham, Joset Jackson, Libby La Goe, Marge Naughton, Candye Treleaven and Mary Tyszkiewicz.

B.A. STOUT
Agricultural Engineering Department
Michigan State University
East Lansing, Michigan

C.A. MYERS, Specialist
Agricultural Engineering Department
Michigan State University
East Lansing, Michigan

A. HURAND, Ingénieur du génie rural, des
    eaux et des forêts
Centre national d'études et d'expérimentation
    de machinisme agricole
Antony, France

L.W. FAIDLEY, Agricultural Engineer
Agricultural Engineering Service
Food and Agriculture Organization

TABLE OF CONTENTS

LIST OF ILLUSTRATIONS

LIST OF TABLES

PREFACE

It is not only misleading but incorrect to speak of "energy shortage". The sun is an infinite energy source. The energy potential of nuclear reactions is also unlimited. Coal supplies should last for several hundred years and petroleum for some decades. The energy "crisis" is thus not one of supply but, rather, of overdependence on petroleum, which is not uniformly distributed throughout the world.

Economic and political reactions to the realization that petroleum supplies are finite have created the "crisis". Consequently, life styles based on extravagant use of cheap petroleum are now being challenged.

The petroleum shortage has triggered an avalanche of studies and reports and has stimulated a re-evaluation of energy alternatives. Priorities for optimum use of valuable petroleum supplies will be needed. Our challenge is to assess the energy situation, to emphasize effective use and elimination of waste and to move diligently ahead with programmes to develop alternatives to petroleum uses in agriculture and the food industry.

Although considerable commercial energy is required to obtain high land and labour productivity, studies show that agriculture is responsible for only a small part of the world energy consumption. Production agriculture used an estimated 3.5% of the world total in 1972. Developed countries used slightly less than the world average, and in the developing countries the proportion was somewhat higher: 4.8%. Continuing expansion of modern agricultural inputs will therefore have only a small effect on world commercial energy demands.

Studies further show that man's well-being is enhanced by his ability to control and use energy effectively. Thus a major consideration of development planners and a goal for researchers in technical fields is to supplement man's limited energy capacity with as much energy as he can afford and use effectively. Environmental and social consequences must be considered as well.

This is not intended to imply that the developing countries should or could follow the lead of the petroleum-gobbling industrial nations. In fact, petroleum price increases have placed a great economic burden on the developing countries and on some developed countries as well. Because petroleum supplies are dwindling, other energy bases must be developed and exploited within the prevailing economic and political framework. The alternative is stagnation of the world economy and a reduced standard of living for mankind.

The purpose of this book is to put energy and the food system in perspective and to provide a manual for persons concerned with maintaining or increasing world food supplies. Energy resources are tabulated, and the principles and efficiencies of various conversion processes are considered, with detailed discussions of their applications in the food system.

In addition, a conceptual outline of the energy flow in the food system and the energy requirements for each operation in production, processing and delivery are presented.

The energy implications of providing an adequate diet for the expanding world population as well as alternative energy sources for the future are discussed.

A major problem in preparing the manuscript was the lack of data from or relating to the developing countries, for which most of the available information comes from the developed countries. More comprehensive data on energy flow in the food systems of the developing countries must be collected in order to develop adequate alternatives for supplying food that are less dependent on expensive and possibly insecure petroleum supplies.

1. WORLD ENERGY RESOURCES

Introduction

Energy is the capacity to do work, that is, to cause a body to move by exerting a force on it. Scientists describe the rate at which work is performed in units of power. The history of civilization is largely a story of man's progress in harnessing energy by converting it into a useful form. Table 1 lists some of the relevant milestones in man's development.

A high level of agricultural or industrial productivity can be attained by a community only when it is able to harness energy equal to many times the total muscular capabilities of its members; thus the poorer areas of the world must develop their energy resources before they can offer the fuller lives available to people in the more advanced areas (274).

More energy is needed for two reasons: (1) the world population is growing at an exponential rate (Fig. 1); and (2) energy use per caput is steadily increasing (Fig. 2). Because petroleum and certain other energy sources (natural gas, coal, etc.) are limited, the energy base for continued development must shift to renewable or practically infinite forms such as solar or nuclear energy.

Much of the world's food is produced, transported and processed by human and animal power, both of which may be categorized as renewable forms of energy. Humans and animals, although limited in individual energy capacity, represent a large energy resource because of their sheer numbers. For example: if the 4 000 million humans on earth were to exert 75 watts for 8 hours a day 250 days per year, they would annually produce 600×10^9 kilowatt-hours (2×10^{18} joules); applying the same logic, if even 10% of the world's 1 500 million head of potential draught animals (horses, mules, asses, cattle, buffaloes, camels) were used at an average output of 500 watts for 8 hours a day, their total energy production would be 112×10^9 kilowatt-hours (4×10^{17} joules) in a 250-day work year (75, 231).

Humans and animals are chemical converters of food energy into

TABLE 1. – CHRONOLOGY OF MILESTONES IN THE HISTORY OF MAN AND ENERGY

1700000 B.C.	First Ice Age begins. Several varieties of erect, manlike primates exist.
Before 500000	Man begins to use fire.
9000	Beginnings of agriculture.
7000	First sickles (found in Palestine).
6000	Domestication of goats, pigs, sheep, cattle, oxen.
4000	Domestication of the horse.
3500	Wheel invented (probably in Mesopotamia).
1000	Beginning of iron technology.
300	Waterwheels in Greece.
200	Modern harness invented in China.
A.D. 650	First windmills. Modern horse harness.
852	Coal burned in an English monastery.
1239	Coal used as fuel by smiths and brewers.
1300	First coal used in home heating.
1600	Versailles waterworks produce 56-kilowatt power.
1606	First known experimental steam engine.
1673	Huygens builds internal-combustion engine run on gunpowder.
1690	Papin designs the first piston engine.
1693	Leibnitz states the law of conservation of potential and kinetic energy.
1712	Newcomen builds first steam-pumping engine.
1765	Modern steam engine conceived by Watt.
1789	Coulomb's work in electrostatics.
1820–1860	Work of Oersted, Ampere, Faraday, and Maxwell in electricity. Principles of thermodynamics worked out by Carnot and Clausius.
1857	First oil well drilled, Pennsylvania, U.S.A.
1876	Otto designs four-stroke internal-combustion engine.
1882	First incandescent lighting, New York.
1896	Becquerel discovers radioactivity.
1903	First flight of the Wright brothers.
1926	Goddard fires first rockets with liquid propellant.
1941	First jet flight.
1942	Fermi starts first atomic reactor in Chicago.
1945	First nuclear explosion, New Mexico, U.S.A.
1952	First hydrogen bomb explosion.
1957	First artificial satellite (Sputnik I). First nuclear power plant, U.S.A.
1969	Man lands on the moon.
1973	Arab oil embargo; quadrupling of crude oil prices.

SOURCE: (236)

FIGURE 1
World population since 1650 has been growing exponentially at an increasing rate. The estimated population in 1970 was already slightly higher than the United Nations projection (made in 1958) shown here. The present world population growth rate is about 2.1% per year, corresponding to a doubling time of thirty-three years (226).

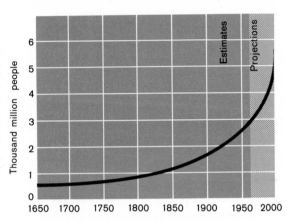

mechanical energy, or work. However, man's power capacity is extremely limited. The value of man's work is insufficient to achieve a reasonable standard of living. Even when man's energy capacity is supplemented by draught animals (75 + 500 watts), the result is often a bare subsistence standard of living. Tables 2–4 summarize the draught power capabilities of various animals.

In the past, inexpensive petroleum and other energy supplies, combined with rising wage rates, made mechanization economically attrac-

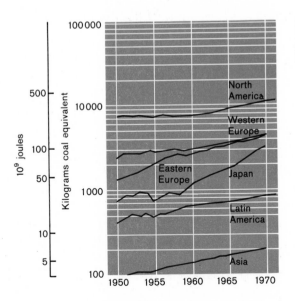

FIGURE 2
Regional per caput energy consumption, 1950–70 (26).

tive to farmers. Energy price increases and uncertain fuel supplies may eventually reverse this trend, but a recent study indicates that a doubling of energy prices in the U.S.A. would result in only a 13% increase in food prices (63).

Some advocate a moratorium on mechanization and a return to a more labour-intensive food production system in view of worldwide unemployment and underemployment (172). A more rational policy would be the development of other energy sources and the technology to apply them.

As Figure 2 has a logarithmic vertical axis, a straight line represents an exponential increase in energy consumption. In the developed

TABLE 2. – ESTIMATED NORMAL DRAUGHT POWER OF VARIOUS ANIMALS

Draught animal	Weight (kg)	Average speed of work (m/h)	Power developed	
			(kW)	(HP)
Light horse	400–700	3 600	0.75	1.00
Bullock	500–900	2 160–3 240	0.56	0.75
Buffalo	400–900	2 880–3 240	0.56	0.75
Cow	400–600	2 520	0.34	0.45
Mule	350–500	3 240–3 600	0.52	0.70
Donkey	200–300	2 520	0.26	0.35

SOURCE: (155)

TABLE 3. – TRIALS OF BULLOCK DRAUGHT PERFORMANCE

Type	Weight (kg)	Mean effort (kg)	Speed (km/h)	Power		Hours of work per day
				(kW)	(HP)	
Pair of White Peul Zebu bullocks (Saria)	790	110	2.3	0.69	0.93	4.05
Pair of N'Dama bullocks (Farako-Ba)	720	100	2.9	0.79	1.06	2.15
Pair of Renitelo bullocks (Kianjasoa)	1 110	150	2.9	1.20	1.61	3.40

SOURCE: (37)

countries, which started from a much higher base in 1950, the increase in consumption has been vastly greater than in the developing countries of Asia and Latin America. Figure 3 depicts the relative increases in per caput energy consumption in developed and developing countries (87).

Figures 4 and 5 show the wide range of per caput energy consumption in selected developed and developing countries. The U.S.A. with only 6% of the world's population consumes about 35% of the energy (148).

The disparity in the energy consumption of the rich and the poor nations is illustrated in another way in Figure 6. If the current ratio

TABLE 4. – POWER OF DRAUGHT ANIMALS OVER LONG PERIODS: PLOUGHING AND TILLAGE

Type and number	Weight (kg)	Average effort (kg)	Speed (km/ h)	Power		Daily working hours		Duration of trial[a] (days)
				HP	kW	Hours worked	Effective hours	
1 donkey	160	46	—	—	—	6.00	3.00–3.50	14–10
1 pair of N'Dama oxen (Sefa)	657	90	2.2	0.72	0.54	5.50	5.50 [b]	4–4
1 pair of N'Dama oxen (Mirankro)	800	80	2.0	0.59	0.44	5.00	4.00	11–10
1 pair of Madagascar Zebu bullocks [c] (Kianjasoa)	650	80	2.5	0.75	0.56	4.75	4.75 [b]	3-3
2 pair of Madagascar Zebu bullocks	1 300	160	1.8	1.07	0.80	5.75	5.75 [b]	2–2
3 pair of Madagascar Zebu bullocks	1 945	200	1.6	1.18	0.88	5.00	5.00 [b]	2–2
1 pair of half-Brahma oxen (Niadana)	1 060	147	2.4	1.30	0.97	5.10	4.40	11–10

SOURCE: (37)

[a] The first figure indicates the span of the trial and the second the number of days worked. – [b] Effective hours are identical to hours worked as the animals were used morning and evening for 2–3 hours; thus it was not necessary to break these half-sessions with rest periods. – [c] Three pair of Madagascar Zebu bullocks — first two pair in harness and then three pair by adding another pair — ploughing an unprepared field; subsequently, a single pair of the original unit did the harrowing.

between energy consumption and population growth continues for fifty years (from 1970), per caput energy consumption (in coal equivalents) will rise to 54 metric tons (146×10^{10} joules) in the rich countries and to 1.4 metric tons (4×10^{10} joules) in the poor countries. By that time the world population would be 10 500 million — about 1 400 million in the rich countries, 8 500 million in the poor countries

FIGURE 3
Estimated and projected per caput energy consumption in the developed and developing countries, 1900–2000 (273).

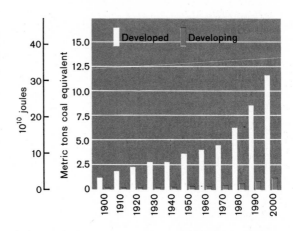

FIGURE 4
Ratio of per caput energy consumption in various countries and regions to the world average (129).

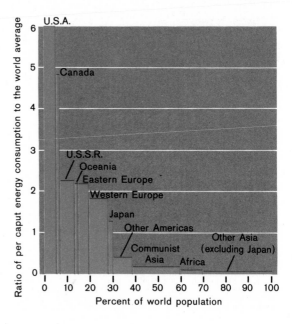

and the rest in the intermediate countries. Energy consumption in the poor countries alone would rise to a level considerably higher than the total consumption of the world today; the world consumption would approach the equivalent of 100 thousand million metric tons of coal (27×10^{20} joules) annually.

Energy consumption density per unit of land area is shown in Table 5. As expected, the densities in large countries such as the U.S.A. and

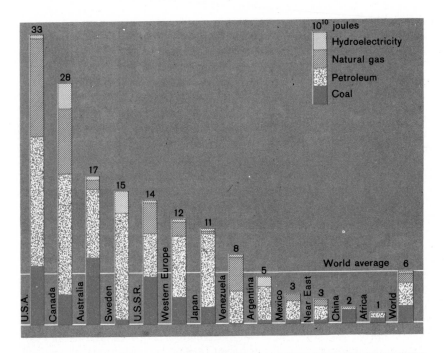

FIGURE 5 (*above*)
Per caput energy consumption (10^{10} joules) in various countries and regions, showing sources, 1973 (40).

FIGURE 6 (*right*)
Population and energy consumption of the richer and poorer nations (26).

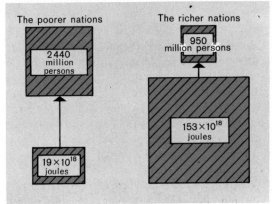

TABLE 5. – ENERGY CONSUMPTION DENSITY IN SELECTED COUNTRIES, 1971

Country	Area (thousand km^2)	Energy consumption density (watts/m^2)
France	573	0.34
Germany, F.R.	246	1.24
Germany, D.R.	108	0.95
Italy	299	0.46
United Kingdom	242	1.21
U.S.S.R.	22 400	0.05
Canada	9 976	0.02
U.S.A.	7 760	0.29
Japan	366	0.89
China	9 561	0.04
India	3 267	0.03
Egypt	1 000	0.01
Argentina	2 776	0.02
Brazil	8 506	0.005

SOURCE: (268)

the U.S.S.R. are lower than in the small European countries. The developing countries rank very low in energy consumption per unit of land.

Each energy source listed in Table 6 can be used directly or indirectly to generate electricity, and most can be used to produce heat in industry and agriculture as well as in the home. However, the cost structure and the location (proximity to consumption centres) often determine the economic attractiveness of the resource. Two further considerations are: (1) the environmental impact, as energy from fuel resources tends to create more environmental problems than energy from nonfuel resources; and (2) the nonrenewability of most fuel resources as opposed to the renewability, with certain limitations, of nonfuel resources.

Whether or not a resource will be developed usually depends on economic factors. All energy resources differ in quantity, quality and, for nonrenewable sources, availability. One that is usually regarded as economically advantageous, such as oil, may not be so attractive in some circumstances. Costs vary considerably, and only close study can determine the economic benefits or drawbacks of a specific resource in any area.

As parts of the globe are still unexplored, knowledge of their energy resources is fragmentary. As energy price levels change and technology improves, sources which have been explored and classified as uneconomical may be reassessed (183); this applies to coal, hydroelectric power, and oil from shale and from old oil fields.

Energy sources that are replenished slowly (or not at all) are classified as nonrenewable. These include fossil fuels (e.g., coal, petroleum, natural gas), nuclear fuels (e.g., uranium) and geothermal power. Sources that are replenished more rapidly are termed renewable. These include wood or other organic materials that may be burned or converted into gaseous or liquid fuels. For all practical purposes other energy forms (e.g., solar, wind and water power) are inexhaustible.

Figure 7 shows the earth's natural energy flow. The three original sources are solar, tidal and geothermal energy.

FIGURE 7. Earth's natural energy flow (122).

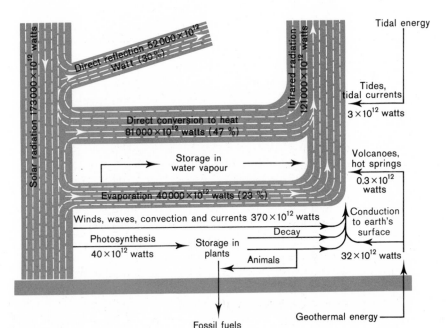

TABLE 6. – ADVANTAGES AND DISADVANTAGES OF VARIOUS ENERGY SOURCES

FUEL RESOURCES

Source	Advantages	Disadvantages
PETROLEUM Essentially a complex mixture of hydrocarbons with small amounts of other substances; recovered from onshore and offshore fields, tar sands and oil shale; also found in deep sea.	Abundant and accessible (world production in 1970 represented only 3.5% of proven world reserves). Deposits are widespread in sedimentary areas. Highly versatile: high-grade fuel is obtained by refining and processing; petroleum and its by-products are used for transportation, heating, lighting, cooling, lubricating, medical products, animal protein, fertilizer, etc.	Nonrenewable. Requires considerable capital investment. Causes atmospheric pollution through combustion. Cost of production from tar sands and oil shale is higher than from conventional sources. Offshore exploration and drilling is more expensive than on land. Extracting oil from deep-sea areas beyond the continental shelf involves technological and legal problems. There are growing difficulties in maintaining an equilibrium between supply and demand.
NATURAL GAS A combustible gaseous mixture that in gas fields ("non-associated gas") contains largely methane and in wet state with petroleum ("associated gas") contains other hydrocarbons. Found in natural gas fields, in coal mines, in geopressure zones; obtained from bio-digesters and as a by-product of coke making.	Relatively cheap and abundant. Relatively clean, virtually sulphur-free (except for sour gas). Versatile: used as a raw material for petrochemicals.	Nonrenewable except when produced from organic waste or algae. Expensive to transport when liquified. Risky to handle because of vapour clouds and danger of fire. Causes some atmospheric pollution when used in power plants. Not yet used in areas where no large markets are available to justify high cost of transport.
COAL A solid combustible mineral substance	Very abundant (world production of coal	Nonrenewable.

as anthracite, bituminous, subbituminous, and lignite.	High-grade coal contains 70-80% of the energy per unit weight of oil. Some kinds of coal are low in sulphur. Lignite can be used to produce a high-grade smokeless fuel through the briquetting process.	productive land unless remedial work is undertaken, which may be expensive. Causes atmospheric pollution through combustion releasing carbon dioxide, sulphur dioxide and fly ash. More expensive to transport than oil.
PEAT Compressed and carbonized vegetation found in bogs. Used as fuel when dried or briquetted. Gas can be obtained from by-products.	Moderately widespread in many parts of the world. Can be used locally for domestic purposes and for electricity generation. Low cost if no transport is involved.	Can be more costly than coal to produce on a commercial basis. Nonrenewable.
NUCLEAR FISSION Splitting atoms of heavy elements such as uranium and thorium results in the release of enormous quantities of energy. Plutonium is produced in nuclear reactors. Uranium is found in rocks and sea water; also as a by-product of minerals, such as gold, phosphate, oil shale.	Uranium fairly widespread in nature. Powerful energy source: uranium releases 20 000 times as much heat as the equivalent weight of coal. Could provide unlimited energy resources through nuclear breeder reactors. Power plants can generate electricity. Heat could possibly be used for desalination of water.	Installation of nuclear plants needs much capital, and suitable sites are difficult to find. All existing types of reactors consume more fissionable material than they produce. Pollution: thermal waste (heated water dumped into river or sea) threatens aquatic life; radioactive waste is hazardous to health; plutonium used in breeder reactors is a dangerous poison. Storage facilities of radioactive substances have a short life in relation to the life of the radioactive materials. Material can be stolen or otherwise diverted for use in nuclear weapons. Control of use of atomic energy raises complex political issues.
WASTE PRODUCTS Agricultural and municipal wastes provide steam when burned; animal waste can be	Easily obtained and renewable. Can be used for the generation of electricity.	Organic municipal waste produces low-grade fuel.

TABLE 6. – ADVANTAGES AND DISADVANTAGES OF VARIOUS ENERGY SOURCES (*continued*)

Source	Advantages	Disadvantages
dried and used directly as a fuel or converted to methane by fermentation and to oil or gas by methods of decomposition.	Can be reprocessed to produce cattle feed. Solves problems of waste disposal and of related environmental pollution.	Large-scale collection of agricultural organic waste could be costly. Technical problems are still to be solved (e.g., how to handle nonorganic solids). Can only complement other sources of energy.
WOOD A traditional source of energy.	Provides heat for domestic purposes. Methanol can be produced from wood. Renewable. Less polluting than other fuels.	Provides less heat per unit of weight than other fuels, such as coal and oil. Inefficient conversion causes smoke pollution. Other industrial uses, such as construction and paper production, may yield a higher return than its use for energy. Forests are far from industrial centres.

NONFUEL RESOURCES

Source	Advantages	Disadvantages
HYDROPOWER [1] Waterpower used to supply energy.	Clean method of electricity production. Can be a cheap source of electricity.	May involve high initial construction cost. Growing shortage of natural sites. Damming the water may cause changes in the environment, backwater sedimentation and rapid silting.
GEOTHERMAL Energy supplied from the heat of the earth's interior, hot springs, hot rocks.	Abundant. Can generate electricity and provide heat for domestic, agricultural and industrial purposes.	Found principally in areas of tectonic activity. Environmental pollution possible:

	tively small power units. Can provide base load of 8 000 hours per year, not subject to seasonal variations. Where district heating or greenhouse heating is required, geothermal heat can be produced at very low cost.	into the field; thermal pollution may be created when used to generate electricity. Hot water and steam must be used close to source. Technology for obtaining energy from geo-pressure zones and hot rocks not yet developed.
TIDAL Power generated from the flow of tides.	Nonpolluting. Renewable.	Possible only in areas where difference in tide levels is high enough to generate electricity. Output is intermittent and depends on tide cycles. Installations are complicated and costly.
WIND [1] Power from force of wind.	Traditionally used in many rural areas (e.g., for pumping water, turning millstones). Nonpolluting. Small wind generators can supply electric energy in isolated regions.	Variation in energy output according to duration and force of wind. Storage of electricity when wind velocity changes is expensive. For large-scale production suitable sites with adequate wind power are hard to find. Can only be complementary to other sources of energy.
SOLAR ENERGY Sunlight affects rains, winds and ocean currents; provides energy for plant and animal life cycles through photosynthesis.	Direct use of heat for water and space heating, cooking, drying crops, desalination of water, evaporation to produce salt. Renewable, diffuse and inexhaustible.	Limited hours of sunlight and variation of solar intensity. Solar collectors can only provide low-grade heat on a small scale.

[1] Function of solar energy.

TABLE 6. – ADVANTAGES AND DISADVANTAGES OF VARIOUS ENERGY SOURCES (*continued*)

Source	Advantages	Disadvantages
SOLAR ENERGY (*continued*)	Nonpolluting and safe. Considerable potential for space cooling; water pumping; conversion of sunlight to electricity through solar cells and use of photosensitive materials such as silicon; possible use for extraction of hydrogen from water.	Technical difficulties in using on a large scale: conversion and collection techniques must be improved; back-up plant is required to ensure continuity of supply from such an intermittent source. Power systems entail high initial costs. At present can only supplement other sources of energy.

POSSIBLE FUTURE RESOURCES

Source	Advantages	Disadvantages
SOLAR ENERGY FROM OUTER SPACE Solar cells in space satellites generate electricity and send it by microwaves back to receiving stations on earth.	Sunlight is inexhaustible. Needs no storage system.	System likely to be extremely costly. Could produce thermal pollution if used on a large scale. Legal status of use of energy will have to be settled.
NUCLEAR FUSION The union of atomic nuclei of light chemical elements to form nuclei of heavier elements, resulting in the release of 'enormous quantities of energy. Tritium and deuterium (isotopes of hydrogen) are the probable fuels to be used in a fusion reactor. Deuterium and tritium present in sea water.	Virtually inexhaustible. Likely to be safe. Pollution-free, no release of carbon dioxide. Might breed its own fuel and burn it immediately.	Fusion reactors are not yet scientifically feasible.
SEA-THERMAL [1] Difference in temperature between warm	Renewable.	Energy produced is expensive

	duce electricity by means of a heat engine operating across the temperature differential.	Benefits from simultaneous production of electricity and desalinated water. Fish farming would be possible with nutritious sea-water pumped from depths.	
WAVES [1]	Rolling motion creates energy for potential use.	Nonpolluting. Safe.	Intermittent except where constant waves prevail. Pilot plants are being constructed. Adequate wave power distributed unevenly along the coastlines of the world. Probably not practicable for large-scale supplying of energy.
OCEAN CURRENTS [1]	Speed and motion of current used to generate power.	Nonpolluting. Safe.	Not concentrated. Systems for electricity generation similar to those developed for tidal power are needed. Not practicable for large-scale supplying of energy.
ALGAE (INCLUDING KELP)	Methane is produced when algae are digested by bacteria.	Easy to grow and harvest on land, in freshwater ponds and in ocean areas. Methane could be used to feed fuel cells and produce electricity.	Present kelp yield from the sea is not large in relation to world requirements.
VOLCANOES		Contain large quantities of energy.	Technology to make use of volcanic energy not yet available. Volcanic heat is less than that produced by hot springs.

SOURCE (246)

[1] Function of solar energy.

Renewable or inexhaustible resources

SOLAR

The sun produces energy by converting mass into energy at a rate of millions of tons per second. The total amount of energy striking the outer atmosphere of the earth in a year is 35 000 times the energy used annually by man. The average intensity measured on a plane perpendicular to its path is 1.36 kilowatts per square metre when it reaches the earth's atmosphere. This number is known as the solar constant. Only when the sun is directly overhead is energy received at the maximum rate. At other times the rate depends on the angle at which the radiation strikes the surface (171).

Much of the sun's energy is screened out by the earth's atmosphere or reradiated into space. Figure 8 shows what happens to the radiation intercepted by the earth. The maximum intensity of solar radiation at the earth's surface, about 1.2 kilowatts per square metre, is encountered only at or near the equator on clear days at noon. Under these ideal conditions the total energy received is from 6 to 8 kilowatt-hours (22 × 10⁶ to 29 × 10⁶ joules) per square metre per day (226).

FIGURE 8
Interception of solar radiation by the earth (226).

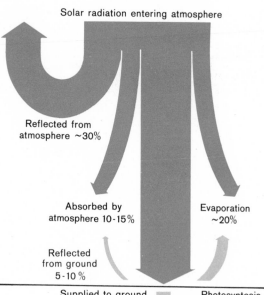

Solar radiation entering atmosphere

Reflected from atmosphere ~30%

Absorbed by atmosphere 10-15%

Evaporation ~20%

Reflected from ground 5-10%

Supplied to ground and reradiated ~30%

Photosyntesis, wind, ocean currents, etc. <1%

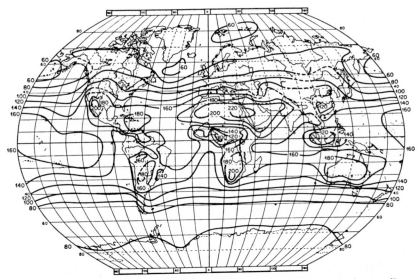

FIGURE 9. Geographical distribution of the solar radiation received annually at the ground. The contours are labelled in kilolangleys, or kilocalories per square centimetre (214).

The geographical distribution of solar radiation received at the earth's surface is shown in Figure 9. At a radiation intensity of 120 kilocalories (5×10^5 joules) per square centimetre per year the solar energy falling on about 59×10^5 hectares (59 000 square kilometres) would meet the total worldwide energy requirement (3×10^{20} joules).

Solar energy is not, of course, available continuously because of the day/night cycle and cloud cover and because its intensity varies according to season, geographical location and other factors. Research is under way for the development of practical solar heat collectors, heat storage and distribution systems to supplement or replace nonrenewable energy forms. (Solar energy systems are discussed in detail on pages 87–117.)

WIND

Air temperature differences cause winds. The main energy force driving the flow of air is the temperature difference between the tropic and the polar regions. In the tropical regions, which are more perpendicular to the sun and exposed to sunlight for a longer time, the much warmer air expands, causing it to rise and push up the top of the atmosphere. The air over the tropics thus rises higher above the earth's surface and flows toward the poles. Some of the warm air descends at the mid-latitudes, and the rest goes to the poles.

In the polar regions the air is cooled and tends to contract into a heavy, dense mass until its weight causes it to flow from the polar regions toward the tropics. Because of the earth's rotation, the air also circulates in other spirals and bands, but it is basically the difference between tropical and polar temperatures that causes the winds and their motions (180).

As wind results from sunlight, it is often considered a type of solar energy. Because the technology for capturing wind energy differs markedly from that of solar heat collectors, wind as an energy source is here discussed separately.

The worldwide average wind velocity in the lower atmosphere is about 9 metres per second (216). This is equivalent to about one half kilowatt per square metre of "windmill" area perpendicular to the flow. The power that may be extracted from the wind varies as the cube of the velocity. The power provided by a 16 kilometre per hour wind is about 54 watts per square metre; at twice that velocity, 32 kilometres per hour, it is 430 watts per square metre ($2^3 = 8$ times the power).

The total worldwide wind power potential is estimated at 2×10^{10} kilowatts (226). However, as wind energy is very dilute — that is, the energy density is low — a large machine is needed to capture a small amount of energy. The cost effectiveness of wind machines must be scrutinized carefully.

Another problem is that winds are erratic. Wind energy can only be used intermittently (when the wind blows) until some type of energy storage system is devised. Nevertheless, extensive research is under way on capturing some of the wind's energy and putting it to work for mankind. (See Chapter 3 for detailed information about wind energy systems.)

ORGANIC ENERGY RESOURCES

Solar energy can also be put to use through photosynthesis and the growth of organic matter (e.g., biomass). From the beginnings of civilization until early in the nineteenth century most energy needs were met by recurrent sources, mainly wood (Fig. 10). Wood has gradually been replaced as a major fuel source in developed countries, but many developing countries still rely heavily on it as fuel.

Dry organic matter, such as plant residue, normally contains 13.9–16.2×10^6 joules per kilogram. Energy can be extracted from dry organic matter by numerous processes, such as combustion in excess air or in a controlled atmosphere, pyrolysis and hydrogasification. Wet organic residues, such as animal excrement, can be converted

FIGURE 10
Contribution of wood fuel
to the energy systems of
Argentina, Germany, Ja-
pan, U.S.A. and Russia–
U.S.S.R., 1800–1950.

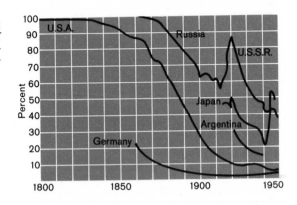

into convenient gaseous fuels by anaerobic fermentation, in which
selected bacteria break down cellulose to form combustible methane.

The present annual production of biomass on the land area of the
world is 10^{11} metric tons of dry matter (1). This is an energy equivalent
of six times the current worldwide energy consumption. (The collection and use of organic residues as an energy source are discussed in
detail on pages 188–206.)

HYDROPOWER

The world's estimated waterpower potential is about 2.8×10^9
kilowatts (Table 7). According to the *United Nations Statistical Yearbook* (257) only about 10% (278×10^6 kilowatts) of the world's waterpower capacity was used in 1970. The developing countries in South
America, Africa and Asia have exploited only a minute fraction of
their hydropower; thus its potential as a future energy resource is
significant.

Nonrenewable resources

Table 8 summarizes the world's nonrenewable energy reserves (266).
Figure 11 shows that 49% of all recoverable reserves are located in
North America. Table 9 and Figure 12 give world energy consump-
tion by region. From 1960 to 1972, it increased by 4.9% annually.
The projected annual rate of increase from 1972 to 1990 is 3.3% (266).

Table 10 and Figure 13 show the world's energy consumption by
source. For 1972–90 the projected annual increases are: for coal,
1.8%; for petroleum, 2.4%; for natural gas, 2.9%; for hydropower
and geothermal power, 2.1%; and for nuclear power, 23.6%. In

TABLE 7. – WORLD WATERPOWER CAPACITY, 1970

Region	Potential (10⁶ kW)	Percent of total	Developed (10⁶ kW)	Percent developed
North America	313	11	88	28
South America	577	20	13	2
Western Europe	158	6	93	59
Africa	780	27	6	1
Near East	21	1	2	8
Southeast Asia	455	16	8	2
Far East	42	1	21	50
Australia and New Zealand	45	2	7	16
U.S.S.R., Eastern Europe and China	466	16	40 [a]	9 [a]
TOTAL	2 857	100	278	10

SOURCES: (120, 243, 257)

[a] Excluding China.

FIGURE 11
World recoverable energy reserves (266).

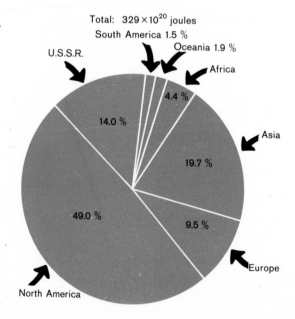

TABLE 8. – MEASURED WORLD NONRENEWABLE ENERGY RESERVES
(10^{18} joules)

Area	Fossil fuels				Uranium (non-breeders)	Total
	Solid fuels	Crude oil	Natural gas	Oil shale and tar sands		
Africa	382	556	213	86	209	1 446
Asia (excluding U.S.S.R.)	2 750	2 330	456	918	3	6 457
Europe (excluding U.S.S.R.)	2 720	60	162	123	49	3 114
U.S.S.R.	3 510	352	610	147	Unknown	>4 619
North America .	5 350	318	402	9 610 [a]	446	16 126
South America .	53	329	64	25	13	484
Oceania	485	10	26	10	105	636
TOTAL	15 250	3 955	1 933	10 919	>825	>32 882

SOURCE: (266)

[a] According to the U.S. Bureau of Mines, the North American oil shale and tar sand reserves may be drastically overstated; development of most of these reserves is uneconomic at present.

TABLE 9. – WORLD ENERGY CONSUMPTION BY REGION, 1960–90
(10^{18} joules)

Region	1960	1965	1970	1972	1980	1985	1990
U.S.A.	47	56	71	76	91	109	129
Western Europe	28	36	49	52	66	79	92
Japan	4	7	13	14	22	28	36
U.S.S.R., Eastern Europe and China	41	48	61	67	86	99	115
Rest of world	19	25	35	38	48	55	64
TOTAL	139	172	229	247	313	370	436

SOURCE: (266)

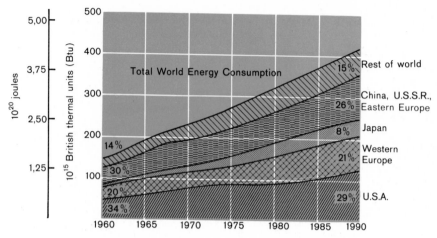

FIGURE 12. World energy consumption by region, 1960–90 (266).

1990, total world energy consumption is projected as: petroleum, 40%; coal, 22%; natural gas, 19%; hydropower and geothermal power, 5%; and nuclear power, 15% (266). Note that world energy consumption in 1976 represented less than 1% of the world's energy reserves.

Figure 14 shows Hubbert's projections (121) for fossil fuel production and resources. The production cycle for any finite resource begins at zero, reaches a maximum and finally declines once more to zero. Because the history of production to the present is known, the curve for the rest of the cycle can be extrapolated.

TABLE 10. – WORLD ENERGY CONSUMPTION BY SOURCE, 1960–90
(10^{18} joules)

Energy source	1960	1965	1970	1972	1980	1985	1990
Coal	65	66	71	70	84	90	97
Petroleum	48	68	102	113	140	155	174
Natural gas	19	28	43	48	60	71	81
Hydropower and geothermal power	7	10	12	14	16	18	20
Nuclear		<1	<1	1	13	36	67
TOTAL	139	173	229	246	313	370	439

SOURCE: (266)

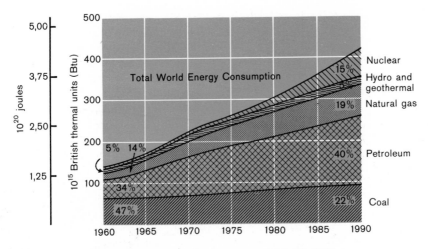

FIGURE 13.   World energy consumption by source, 1960–90 (266).

FIGURE 14.   Complete production cycle of an exhaustible resource (119).

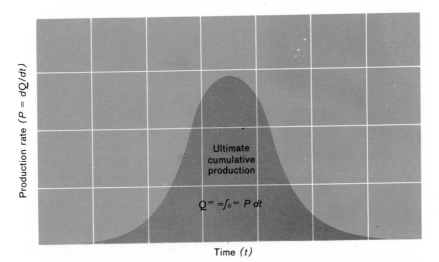

Two different estimates of world petroleum production and potential reserves are given in Figure 15. Bell-shaped curves indicate projected use if the rate of use is proportional to the remaining amount of the resource. Figure 16 presents a similar projection for world coal production.

In Figure 17 each country or area is made proportional to its oil and coal reserves. Not surprisingly, the Near East dominates the oil map and the U.S.A. and the U.S.S.R. loom large in coal (180).

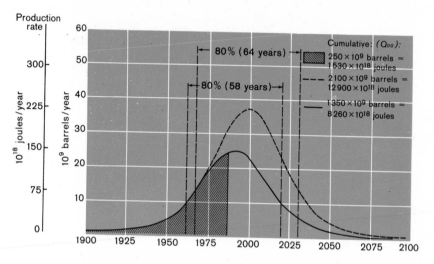
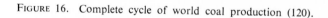

FIGURE 15. Complete cycle of world crude petroleum production (120).

FIGURE 16. Complete cycle of world coal production (120).

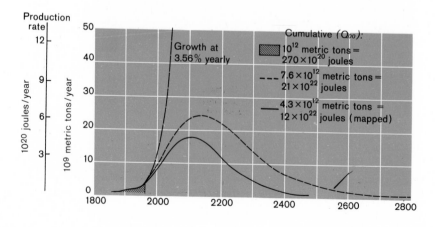

FIGURE 17. World distribution of petroleum and coal reserves. The percentages are shares of the world's known extractable reserves in each area. The size of each country or region is proportionate to its share of reserves (180).

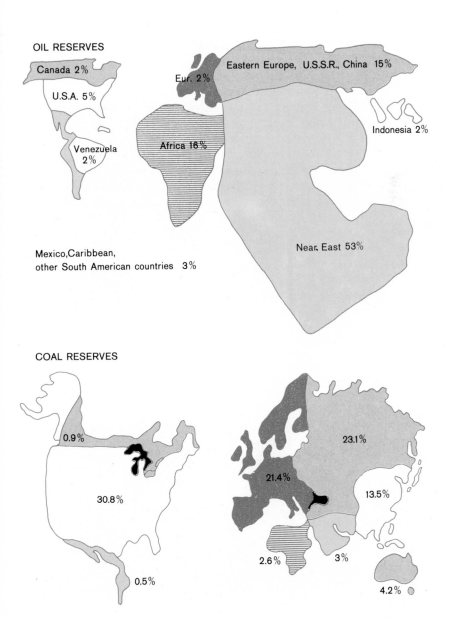

The bulk of oil in international trade is produced in the Near East, Nigeria and Venezuela. Figure 18 shows oil movements by sea between the various parts of the world (23).

The generation of electricity is a vital and rapidly growing conversion technology, permitting the substitution of plentiful nonrenewable resources (coal, nuclear) for others in short supply (petroleum, natural gas). World energy inputs in the electricity sector are shown in Table 11 and Figure 19. These inputs increased 7.6% annually between 1960 and 1972, and they are projected to increase 6.4% annually between 1972 and 1990.

Nuclear energy inputs in the electricity sector are expected to increase 23.6% annually from 1972 to 1990. By 1990 nuclear energy may constitute 37% of the world's energy inputs for the generation of electricity (226). The primary reserves of uranium (U_3O_8) are in North America, Australia, southern Africa and Sweden, with limited amounts in the developing countries.

In 1965 there were 66 nuclear-powered electric generating stations operating in nine countries with a total capacity of just over 7 000 megawatts (1 MWe $= 10^6$ W). By the end of 1975 more than 180 stations were operating in nineteen countries with a total capacity of over 70 000 megawatts (255).

The future role of nuclear power is projected in Table 12; however, unpredictable developments in safety and nuclear waste handling make these projections uncertain (see pages 180–182).

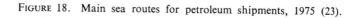

FIGURE 18. Main sea routes for petroleum shipments, 1975 (23).

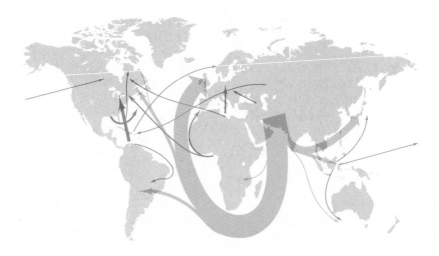

TABLE 11. – WORLD ENERGY INPUTS FOR ELECTRICITY, 1960-90

Year	Fossil fuels	Hydropower and geothermal	Nuclear	Total
	.... 10^{18} joules			
1960	17	7	—	24
1965	26	10	<1	<37
1970	39	12	<1	<52
1972	44	14	1	59
1980	67	16	13	96
1985	77	18	36	131
1990	92	20	67	179

SOURCE: (266)

FIGURE 19. World energy inputs for electricity, 1960–90 (266).

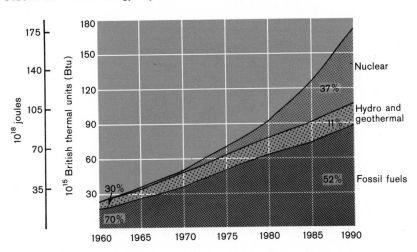

TABLE 12. – OECD/IAEA ESTIMATED NUCLEAR POWER GROWTH

Region	Installed nuclear capacity at end of year				
	1975	1980	1985	1990	2000
	.... Thousand megawatts electrical				
OECD Europe	18.1	77	210	380	810
North America	40.6	89	223	430	1 120
Japan	7.0	17	47	90	170
Developing countries	1.1	9	46	110	400
Countries with centrally planned economies	8.3	38	104	250	640
TOTAL	75.1	230	630	1 260	3 140

SOURCE: (255)

Commercial and noncommercial resources

Energy forms such as petroleum and coal sold through normal market channels are referred to as commercial energy. Many developing countries are reviewing their energy policies and analysing their commercial energy resources (163, 172). A comprehensive example is the 1974 report of India's fuel and power sector (171). In the developing countries noncommercial energy is a major category, consisting of fuels used directly without any conversion technology and without entering formal marketing channels. Firewood is the most important source of noncommercial energy, but crop residues and animal dung are also widely used. (See pages 188–206 for a detailed discussion of organic energy resources and the technology for utilization.)

Energy consumption in various sectors

The exponential curve of increasing energy consumption also applies to the stages of man's development as demonstrated in Figure 20. As long as the energy available to man depended on the food he ate, the rate of consumption was 2 000 kilocalories per day, which the domestication of fire may have raised to 4 000 kilocalories. In

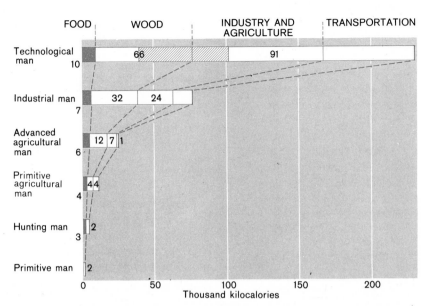

FIGURE 20. Daily energy consumption per caput calculated for six stages of man's development (the accuracy decreases with antiquity). Primitive man (East Africa, about 1 000 000 years ago) had only the energy from the food he ate. Hunting man (Europe, about 100 000 years ago) consumed more food and burned wood for heat and cooking. Primitive agricultural man (Fertile Crescent, 5000 B.C.) grew crops and used animal energy. Advanced agricultural man (northwestern Europe, A.D. 1400) used coal for heating, water and wind power and animal transport. Industrial man (England, 1875) used steam engines. In 1970 technological man (U.S.A.) consumed 230 000 kilocalories daily, much of it in the form of electricity (hatched area).

a primitive agricultural society with some domestic animals the rate of consumption rose to about 12 000 kilocalories or perhaps double that amount in more advanced farming societies.

At the height of the low-technology industrial revolution (1850–70), per caput daily energy consumption reached 70 000 kilocalories in England, Germany and the U.S.A. The succeeding high-technology revolution, brought about by the central electric-power station and the automobile, enabled the average person to use power in his home and on the road.

Shortly before 1900, per caput energy consumption in the U.S.A. began to rise at an increasing rate to the 1970 figure of about 230 000 kilocalories per day for the country as a whole. Today the industrial regions, with 30% of the world's population, consume 80% of the world's energy (46).

The daily flow of energy by source and use in the U.S.A. in 1970 is depicted in Figure 21. Of the total oil supply of 850×10^{14} joules (636×10^{14} from domestic sources and 214×10^{14} imported), only 61×10^{14} joules were used to generate electricity, while the largest share, 453×10^{14} joules, went to the transportation sector (cars, airplanes, trains, etc.).

In addition, the flow chart shows that the 1970 domestic coal production of 453×10^{14} joules per day in heat equivalent was less than that of oil or gas. The figure also shows that nearly half the energy produced is rejected as waste heat. The transport sector is seen to be the least efficient end user of energy. The generation of electricity is inefficient, too, as almost twice as much energy is rejected in waste heat as is produced in electricity.

Economic considerations

The price of a commodity is normally thought of as a mechanism to allocate supply; however, energy prices are often quite unrelated to energy costs because of taxes and regulated or arbitrary pricing.

Higher oil prices have had a sharp impact on the developing countries, sharply increasing the capital of the major oil exporters while imposing a heavy burden on the others — the oil importers (90, 250, 253). Table 13 on pages 32–33 categorizes the developing countries into exporters and nonexporters of oil, the latter by per caput income levels (207).

FIGURE 21. Daily energy flow ($\times 10^{14}$ joules) in the U.S.A., 1970 (135).

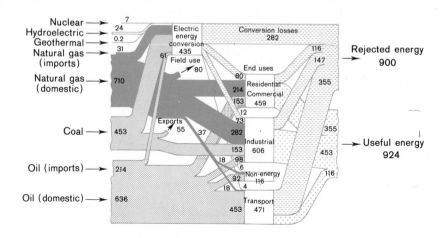

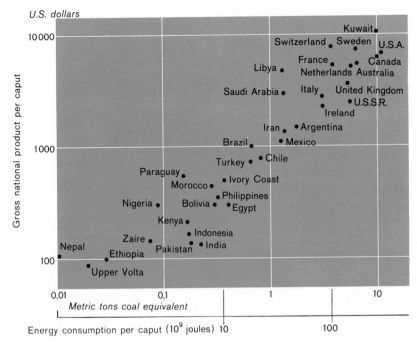

FIGURE 22. Relationship between per caput gross national product and energy consumption in 51 countries, 1975 (126, 249*a*).

Table 14 on page 34 gives the value of exports and imports of selected developing countries and shows that the deficits in trade balances due to oil and manufactured goods imports are not offset by raw material exports.

Many writers have discussed the relationship between social welfare (standard of living) and energy consumption. Gross national product (GNP) has often been plotted as a function of energy consumption, as in Figure 22. Sweden and Switzerland have since surpassed the U.S.A. in GNP per caput with one half to two thirds the latter's per caput energy consumption.[1]

Regardless of the emphatically pointed-out shortcomings of GNP as an index of human betterment, they are often associated (212). A comparison of energy consumption and resultant GNP in the highly developed nations of Europe and North America with those in the developing nations of Africa, Asia and Latin America demonstrates at least a general relationship.

---

[1] For an excellent analysis of energy consumption patterns in the U.S.A. and Sweden, see L. Shipper and A.J. Lichtenberg, Efficient energy use and well-being: the Swedish example. *Science* 194 (4269): 1001–1013 (1976).

TABLE 13. – POPULATIONS AND PER CAPUT INCOMES OF OIL-EXPORTING AND NON-OIL-EXPORTING DEVELOPING COUNTRIES, 1970 [1]

	Population (millions)	Per caput income (U.S. dollars)
Oil-exporting countries		
Algeria	14.33	259 (1969)
Colombia	21.12	366
Egypt	33.33	200
Gabon	0.50	468
Indonesia	117.89	93
Iran	28.66	341
Iraq	9.44	278 (1969)
Kuwait	0.76	3 148
Libya	1.94	1 450
Nigeria	55.07	83 (1969)
Saudi Arabia	7.97	344 (1968)
Syrian Arab Republic	6.25	258
Tunisia	5.14	248
Venezuela	10.40	854
Non-oil-exporting countries		
(*Per caput income above $340*)		
Argentina	23.21	1 000
Brazil	93.39	368
Chile	8.86	614
Lebanon	2.79	521
Mexico	49.09	653
Singapore	2.07	918
Turkey	35.23	352
Uruguay	2.89	787
(*Per caput income between $200 and $340*)		
Bolivia	4.93	202
Dominican Republic	4.06	334
Ecuador	6.09	250
El Salvador	3.53	274
Ghana	8.64	238

[1] Unless otherwise indicated.

TABLE 13. – POPULATIONS AND PER CAPUT INCOMES OF OIL-EXPORTING AND
NON-OIL-EXPORTING DEVELOPING COUNTRIES (*concluded*)

	Population (millions)	Per caput income (U.S. dollars)
Guatemala	5.28	337
Honduras	2.58	256
Ivory Coast	4.31	321
Jordan	2.31	273
Korea, Republic of	31.02	245
Malaysia	10.40	329
Morocco	15.52	212
Paraguay	2.39	230
Peru	13.59	293
Philippines	36.85	228
Senegal	3.93	201
Viet Nam, Republic of	18.83	232
Zambia	4.18	335
(*Per caput income below $200*)		
Afghanistan	17.09	83
Angola	5.58	154 [1968]
Burma	27.58	68 [1969]
Cameroon	5.84	166
Ethiopia	24.63	71
India	539.86	88 [1969]
Kenya	11.23	131
Madagascar	6.75	126
Mali	5.05	50 [1969]
Pakistan	114.18	116 [1969]
Sri Lanka	12.51	161
Sudan	15.70	109
Tanzania	13.27	94
Thailand	34.38	169
Uganda	9.81	127
Upper Volta	5.38	57 [1968]
Yemen, Democratic	7.21	96
Zaire	21.57	87

SOURCE: (207)

TABLE 14. – VALUE OF EXPORTS AND IMPORTS OF SELECTED DEVELOPING COUNTRIES, 1970 (IN 1974 PRICES) [1]
(Million U.S. dollars)

	Oil	Other raw materials	Manufactures and others	Total
Exports				
Developing countries with per caput income:				
Above $340	2 434.0	11 908.4	7 682.7	22 025.1
Between $200 and $340	1 151.3	9 739.2	4 427.5	15 318.0
Below $200	1 419.9	9 935.6	3 650.1	15 005.6
TOTAL	5 005.2	31 583.2	15 760.3	52 348.7
Imports				
Developing countries with per caput income:				
Above $340	5 717.3	7 022.6	14 854.0	27 593.9
Between $200 and $340	3 809.3	4 992.4	8 239.0	17 040.7
Below $200	4 611.7	4 659.2	8 908.9	18 179.8
TOTAL	14 138.3	16 674.2	32 001.9	62 814.4
Balance				
Developing countries with per caput income:				
Above $340	—3 283.3	4 885.8	— 7 171.3	— 5 568.8
Between $200 and $340	—2 658.0	4 746.8	— 3 811.5	— 1 722.7
Below $200	—3 191.8	5 276.4	— 5 258.8	— 3 174.2
TOTAL	—9 133.1	14 909.0	—16 241.6	—10 465.7

SOURCE: (207)

[1] Estimated by using 1974 price ratios and 1970 values.

Figure 23 shows shifts in per caput income as a function of per caput energy consumption in selected countries for 1972 and 1973 (210). Clearly, an essential ingredient of development is increased energy consumption per caput. Moreover, if food supplies are to meet the demand of an expanding world population, energy supplies to the food system must be increased (53).

Environmental considerations

Energy materials, other natural resources and the environment are interwoven in a complex fashion. When social and economic considerations are superimposed, the global interdependence becomes overwhelming. Perturbations introduced anywhere in the ecosystem may have immediate repercussions elsewhere. Table 15 lists some of the concerns of environmentalists. By analysing the items on this list, ranging from urban decay to wilderness preservation, four types of environmental damage can be identified:

(1) threats to human health and safety;

(2) damage to economic resources and to material well-being;

(3) diminished psychological/aesthetic "enjoyment of life";

(4) damage to nonhuman environments — to nature.

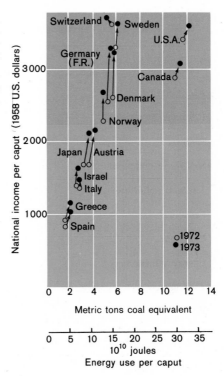

FIGURE 23. Per caput energy use and national income in some important industrial and emerging nations. Note the wide variation in energy use among the nations with the highest per caput incomes. The 1972 and 1973 data were compiled by A. Rosenfeld, Lawrence Berkeley Laboratory, from U.S. and United Nations statistical abstracts.

Examples of direct public health problems are air pollution, water pollution, and excessive noise.

Damage to economic resources may be direct. Air pollution can damage crops and timber as well as paint and stone on buildings. Water pollution increases the cost of purification for household and industrial use. Nitrate fertilizer runoff makes water supplies unfit to drink.

When the economic resource is part of a natural ecosystem, the damage may also be indirect. Often the cause is recognized only after the damage is done, making corrective action difficult or impossible. For example, many commercial fish depend on a food chain consisting of organisms which spend part of their life cycle in coastal marshes and estuaries; thus filled-in marshes and polluted estuaries are damaging to commercial ocean fisheries.

Stream siltation caused by exploitative lumbering damages sport fishing; summer home developments encroach on woodlands, reducing their recreational values (e.g., for hunting and hiking); ocean oil pollution dirties beaches, diminishing our enjoyment of them.

TABLE 15. – PARTIAL LIST OF ENVIRONMENTALISTS' CONCERNS

Water pollution	Visual pollution
Air pollution	Wilderness preservation
Resource depletion	Wildlife preservation
Radioactivity	Urban decay
Solid-waste disposal	Suburban sprawl
Thermal pollution	Agricultural malpractices:
Pesticides	Feedlots — "sewage" disposal
Land use	Fertilizer runoff
Energy utilization	Soil misuse — erosion
Transportation — mass transit	Bad lumbering — clear-cutting
Noise pollution	Oil spills

NOTE: Some items overlap. For example, fertilizer runoff is an example of water pollution, and land use includes aspects of urban decay, suburban sprawl and wilderness preservation. Furthermore, many items are related to one another and thus cannot be dealt with individually.

Examples of damage to nature are unlimited. DDT, which becomes more concentrated as it moves up the food chain (because of its long life and its solubility in fat but not in water), has caused breeding failures in birds, thereby threatening certain species with extinction. Predator control programmes emphasizing poisoning and bounties have nearly eliminated wolves in the continental U.S.A. Lumbering in mature redwood forests has significantly reduced their area. Extensive filling and draining of swamps and marshlands have notably diminished the extent of that type of ecosystem, causing a decrease in some plant and animal species (243).

Climate may also be affected by the heat liberated from human-induced or -controlled energy conversion. It might seem unlikely that man could influence the large-scale energy flows that determine continental and global climates; however, two considerations suggest that the matter should be given more careful thought.

First, about 20% of the land surface of the earth has been drastically altered by man, much of it in preindustrial times. Forests have been converted to grasslands, and overgrazing has turned grasslands into semideserts. Large-scale surface modifications change the reflectivity and therefore the radiative balance.

A reduction in vegetation also alters the distribution of latent and sensible heat. This may bring about a permanent change in the climate of a region, specifically a decline in rainfall. There is evidence that some of the world's deserts were man-made. Such alterations were caused by civilizations using far less power than man does today.

Second, in comparing the power levels of man and nature (Table 16) it has been assumed that meteorological change must be caused by

TABLE 16. – MAN'S POWER USE COMPARED WITH SOLAR POWER LEVEL

Mode of energy flow	Power level (10^{12} watts)
Radiation	173 000
Direct reflection	52 000
Direct conversion to heat	81 000
Evaporation	40 000
Man's power use (1970)	6

SOURCE: (243)

direct assault. As the earth's climate results from a delicate balance among large energy flows, even a small imbalance could lead to considerable changes in the global meteorological system (243).

Man now produces energy at the rate of 500×10^{14} kilocalories per year, which represents 1/20 000 of the total energy received by earth from the sun and 1/5 000 of the total energy received by the earth's land mass. Man's energy use is now increasing by about 5% annually. At this rate, within two hundred years man will be using as much energy as he receives from the sun. Long before this happens, man must come to terms with the global climatological limits on his energy use (277).

In June 1972 a United Nations Conference on the Human Environment was held in Stockholm, Sweden, to bring attention to the increasing evidence of the deleterious effects of man's activities on the natural environment, with serious risks for his survival and well-being, and to propose cooperative action to meet this challenge.

Developing nations are concerned about the environmental aspects of energy development and use (213). In March 1976 the United Nations Economic Commission for Africa convened the Second African Meeting on Energy, in Accra, during which environmental issues were discussed with particular reference to energy development and use (254).

Each country is necessarily planning for its own energy needs in terms of national interest; but it is important that all nations cooperate on analysis of the overall energy issue, strategies to cope with energy problems and recommendations for the future. People everywhere must be made aware of the possible future effects of government decisions on energy development and use, as well as of the need for cooperative effort. The effectiveness of energy-saving technologies will depend on the combined efforts of all nations to deal with this challenge realistically by sharing capital and resources.

2. ENERGY FLOW IN AGRICULTURE

Levels of agricultural development

The earliest human inhabitants of the earth were hunter-gatherers. Under favourable conditions they required at least 1.5 square kilometres to provide food for one person, and in harsher environments as much as 80 to 100 square kilometres (145). Population pressure eventually led man to raise plants and animals.

Shifting cultivation, one of the first agricultural practices developed, is still widely used. At present an estimated 36 million square kilometres of land (about 30% of the world's exploitable soils) are farmed under this system, producing food for about 250 million people (77).

Each hectare under shifting cultivation can provide an all-crop diet for one person (145). Degradation of soil and vegetation usually occurs when the population density exceeds one person per 4 hectares. At higher population densities, shorter fallow periods and eventually annual cropping become necessary.

Annual cropping by traditional methods requires more labour and yields are lower. Animal power can help reduce human labour and provide manure for fertilizer; however, draught animals must be either fed some of the crop or pastured, thereby increasing the land area required per person unless yields per unit of land increase accordingly.

Dramatic increases in crop yields per hectare have been achieved in the developed countries. Although many developing countries have reserves of unused but potentially productive land, most of them will likewise have to meet rising food demands by substantially raising yields on used and new land. Even the clearing of new land, ever less accessible, will require large inputs of commercial energy and often the provision of drainage, irrigation and soil conservation systems. Once the land is under production, additional energy will be needed for maintenance.

Tables 17 and 18 provide estimates of the commercial energy required for rice and maize production at the different levels of technology as well as respective yields. Table 19 lists energy subsidies for various

TABLE 17. – COMMERCIAL ENERGY REQUIRED FOR RICE PRODUCTION BY MODERN, TRANSITIONAL AND TRADITIONAL METHODS WITH RESPECTIVE YIELDS

Inputs:	Modern (U.S.A.)		Transitional (Philippines)		Traditional (Philippines)	
	Quantity/ha	Energy/ha (10^6 joules)	Quantity/ha	Energy/ha (10^6 joules)	Quantity/ha	Energy/ha (10^6 joules)
Machinery	4.2×10^9 joules	4 200	335×10^6 joules	335	173×10^6 joules	173
Fuel	224.7 litres	8 988	40 litres	1 600	—	—
Nitrogen fertilizer	134.4 kg	10 752	31.5 kg	2 520	—	—
Phosphate fertilizer	—	—	—	—	—	—
Potassium fertilizer	67.2 kg	605	—	—	—	—
Seeds	112.0 kg	3 360	110 kg	1 650	107.5	—
Irrigation	683.4 litres	27 336	—	—	—	—
Insecticide	5.6 kg	560	1.5 kg	150	—	—
Herbicide	5.6 kg	560	1.0 kg	100	—	—
Drying	4.6×10^9 joules	4 600	—	—	—	—
Electricity	3.2×10^9 joules	3 200	—	—	—	—
Transport	724×10^6 joules	724	31×10^6 joules	31	—	—
TOTAL		64 885		6 386		173
Yield (kg/ha)	5 800		2 700		1 250	

SOURCE: (81)

TABLE 18. – COMMERCIAL ENERGY REQUIRED FOR MAIZE PRODUCTION BY MODERN AND TRADITIONAL METHODS WITH RESPECTIVE YIELDS

	Modern (U.S.A.)		Traditional (Mexico)	
	Quantity/ha	*Energy/ha (10⁶ joules)*	*Quantity/ha*	*Energy/ha (10⁶ joules)*
Inputs:				
Machinery	4.2×10^9 joules	4 200	173×10^6 joules	173
Fuel	206 litres	8 240	—	—
Nitrogen fertilizer ...	125 kg	10 000	—	—
Phosphate fertilizer ..	34.7 kg	586	—	—
Potassium fertilizer ..	67.2 kg	605	—	—
Seed	20.7 kg	621	10.4 kg	—
Irrigation	351×10^6 joules	351	—	—
Insecticide	1.1 kg	110	—	—
Herbicide	1.1 kg	110	—	—
Drying	$1\ 239 \times 10^6$ joules	1 239	—	—
Electricity	$3\ 248 \times 10^6$ joules	3 248	—	—
Transport	724×10^6 joules	724	—	—
TOTAL		30 034		173
Yield (kg/ha)	5 083		950	

SOURCES: (81, 189)

food crops with different production practices. The only commercial energy input in traditional farming is that needed to produce tools and animal implements. As the traditional farmer uses no commercial fertilizer and plants seed from the previous crop, no commercial energy inputs are involved. However, yields per hectare in traditional farming are low.

In the transition from traditional to modern farming, commercial energy use increases sharply. Primary tillage, usually the first operation mechanized, requires more commercial energy for the production and operation of farm machinery. Improved varieties, often introduced during the transitional phase, require commercial energy for their production and distribution. To help realize yield potentials, soil fertility is improved by using commercial fertilizers and pesticides.

During the transitional phase in rice production in the Philippines about 70% of the commercial energy inputs are for fertilizer, seed production and pesticides and about 30% for farm machinery and fuel. Although Table 17 shows commercial energy inputs for irrigation only under modern methods, supplementary irrigation with mechanically powered pumps will often double yields during the transitional stage.

The commercial energy input for modern rice production in the U.S.A. is ten times that used for transitional production in the Philippines. About 42% of the energy is used for irrigation, 20% for the manufacture and operation of farm machinery, 18% for fertilizers,

TABLE 19. – ENERGY SUBSIDY FOR VARIOUS FOOD CROPS

Kilocalories of energy subsidy per kilocalorie of food output	Type of production
0.02–0.05	Rice production in Indonesia, China and Burma with hand labour supplemented by minimal draught power.
0.05–0.1	Rice production in Thailand. Large-scale cultivation of directly consumed potatoes.
0.1–0.2	Hunting and gathering. Intensive rice production in Europe.
0.2–0.5	Extensive maize cultivation. Intensive potato cultivation. Intensive soybean cultivation.
0.5–0.9	Intensive maize cultivation. Family egg production. Extensive beef production.
About 1	Dairy farming on grassland. Coastal fishing.
2–5	Beef production on grassland. Industrialized egg production.
5–10	Powdered fish proteins.
10–20	Fishing. Livestock raising on feedlots.

SOURCE: (227)

and 7% for crop drying. Modern maize production in the U.S.A. requires only about half the commercial energy that modern rice production does, as little or no irrigation is necessary. Farm machinery is the largest user of commercial energy in modern maize production, accounting for 41% of the total input, followed by fertilizers with 37%.

With modern production methods 1 500 kilograms of petroleum per hectare are needed for rice and 700 kilograms for maize. However, with this commercial energy use, yields of 5.8 metric tons per hectare have been obtained for rice and 5 metric tons per hectare for maize — about five times those obtained with traditional methods. Thus 20–25 people can be fed on an all-grain diet from a single hectare compared with 4–6 people by traditional methods.

The transition to more modern production methods varies from one area of the world to another, particularly the cost of land and labour relative to each other and to agricultural prices (111). These areas can be classified into three groups: (1) where land is inexpensive compared with labour, mechanization is introduced to increase production per worker; (2) conversely, where land is costly and labour inexpensive, biological and chemical inputs, such as plant breeding, chemical pesticides and artificial fertilization, are stressed; and (3) others where high productivity of both labour and land is achieved through a combination of labour-saving mechanization and fertilization.

An example of the first group is Australia, where labour-saving mechanization has increased production per worker, but yields remain low. The second group is illustrated by Egypt, Japan and the Republic of Korea, where biological and chemical inputs have greatly increased yields per hectare, but output per worker remains fairly low. The third group is exemplified by two sets of countries: (*a*) France, the Federal Republic of Germany and the United Kingdom, where labour-saving mechanization and high fertilization have resulted in a fairly high level of land and labour productivity; and (*b*) Canada and the U.S.A., in which a high degree of labour-saving mechanization combined with moderate fertilizer use has achieved about the same yields per hectare as in the first set of countries, but with a considerably higher output per worker.

As most modern agricultural inputs demand commercial energy for their manufacture and application, the commercial energy input per hectare and per agricultural worker is expected to parallel closely the output per hectare and per worker. This is illustrated in Table 20, which compares regional estimates of commercial energy input with cereal output per unit of land and per worker.

There is a close relationship between energy input and cereal output per agricultural worker in all regions. The output per worker less

TABLE 20. – COMMERCIAL ENERGY USE AND CEREAL OUTPUT PER HECTARE AND
PER AGRICULTURAL WORKER, 1972

	Energy/ hectare	Energy/ worker	Output/ hectare	Output/ worker
	.... 10^9 joules		.... Kilograms	
Developed countries	24.8	107.8	3 100	10 508
North America	20.2	555.8	3 457	67 882
Western Europe	27.9	82.4	3 163	5 772
Oceania	10.8	246.8	976	20 746
Other developed countries	19.4	19.1	2 631	2 215
Developing countries ...	2.2	2.2	1 255	877
Africa	0.8	0.8	829	538
Latin America	4.2	8.6	1 440	1 856
Near East	3.8	4.4	1 335	1 386
Far East	1.7	1.4	1 328	781
Centrally planned economies	5.9	6.8	1 744	1 518
Asia	2.4	1.7	1 815	911
Eastern Europe and the U.S.S.R.	9.3	28.5	1 682	4 109
World	7.9	9.9	1 821	1 671

SOURCE: (81)

the input per worker is indicative of the standard of living that the
economy can sustain for that category of workers. The low outputs
of labour-intensive systems are inconsistent, however, with the costs
of the minimum living standards desired. The largest energy input,
556×10^9 joules per worker in North America, corresponds to the
largest cereal output, 67 882 kilograms per worker. The rank of the
other developed regions in use of energy per worker coincides with
that of their output per worker. Similarly, among the developing
regions, Latin America is seen to have had both the largest energy
input per agricultural worker (8.6×10^9 joules) and output per worker

(1 856 kilograms), whereas Africa with the lowest commercial energy input per worker (only 0.8×10^9 joules) had the lowest cereal output per worker (538 kilograms).

There is also a close relationship between energy input and cereal yield per hectare in the developing countries. Excluding the Asian centrally planned economies, in 1972 Latin America led the developing countries in commercial energy applied per hectare (4.2×10^9 joules) and had the highest cereal yields (1 440 kilograms per hectare). The other developing regions also ranked in the same order for energy use per hectare as for output per hectare. Again, the lowest commercial energy input was 0.8×10^9 joules per hectare in Africa, where output was only 829 kilograms per hectare.

In the developed countries, however, the relationship between energy input and cereal output per hectare is less close. North America ranked second for energy input per hectare, but it led all regions for yield. This may be explained in part by the high natural fertility of many North American areas. For the other developed regions, yields corresponded more closely to the use of commercial energy per hectare.

Oceania, however, had lower yields per hectare than any developing region except Africa despite a much higher energy use. This mainly reflected (as is evident in Table 23) the much higher proportion of commercial energy used in Oceania for labour-saving mechanization than for fertilizer, in contrast to the developing regions, where a much higher proportion was used for fertilizer. Eastern Europe and the U.S.S.R. also obtained low yields in relation to energy use per hectare.

Considerable commercial energy is therefore required to obtain high land and labour productivity. For example, 20×10^9 joules of energy per hectare are needed to obtain average cereal yields of over 2.5 tons per hectare. Similarly, an average output of 2.5 tons per agricultural worker seems to demand about the same energy input per worker. Nevertheless, it also appears that in the developing countries even small increases in the present low commercial energy use in agriculture will result in substantial production increases.

Agriculture's share of commercial energy

Clearly, rapid increases in commercial energy inputs are needed for a sufficient gain in agricultural production, especially in developing countries where use of these inputs is very low (80). It has been difficult for most petroleum-importing countries to finance sufficient commercial energy supplies since petroleum prices began to rise in 1973; thus it is important for them to determine how much of their total supply of commercial energy is essential for agricultural production.

Estimates of the 1972 commercial energy use in the main regions of the world are presented for fertilizers, farm machinery, irrigation and pesticides later in this chapter. These are summarized in Table 21 and compared with data on total commercial energy use.

Agriculture is responsible for only a small part of total commercial energy use. In 1972 the world average was estimated at 3.5%, about the level in the developed countries, whereas in the developing countries the proportion was only slightly higher: 4.8%. Agriculture's share of total commercial energy use in the various regions ranged from 1.8% in the "other developed countries" (mainly Israel, Japan and South Africa) and 2.8% in North America to 5.3% in the Far East and 6.4% in the oil-rich Near East. Continued expansion of modern agricultural inputs will thus have only a small effect on total commercial energy demands (72).

Table 21 also shows total energy consumption per caput for the total population and per agricultural worker in different regions of the world. Only in North America and Oceania is the energy available per agricultural worker larger than the average per caput energy consumption of the population. In the developing countries the average commercial energy use per person in the population is five times larger than that used to provide agricultural inputs for each agricultural worker.

Present and future use of commercial energy

Tables 17 and 18 show that farm machinery, chemical fertilizers and, to a much smaller extent, irrigation and pesticides consume most of the commercial energy used in agriculture. The estimated commercial energy use for the production and application of these inputs, already given in Tables 20 and 21, are elaborated for each input category in Tables 22 and 23.

In 1972/73 farm machinery manufacture and operation, the largest user of commercial energy in agriculture, accounted for 51% of the world total and ranged from 8% in the Far East to 73% in Oceania. Chemical fertilizer was second with 45% of the world total and ranged from 26% in Oceania to 84% in the Far East; however, in the developing regions it was in first place.

The manufacture and operation of irrigation equipment and the production and application of pesticides each used only about 2% of the total agricultural commercial energy input in 1972/73. The highest

TABLE 21. – ESTIMATED TOTAL AND AGRICULTURAL USE OF COMMERCIAL ENERGY, 1972 73

	Total use	Agricultural use	Percent used in agriculture	Per caput consumption	Energy per agricultural worker
	.... 10^{15} joules			.... 10^9 joules	
Developed countries	135 678	4 637	3.4	183	107.8
North America	76 933	2 140	2.8	333	555.8
Western Europe	42 912	2 114	4.9	119	82.4
Oceania	2 442	137	5.6	154	246.8
Other developed countries	13 391	246	1.8	99	19.1
Developing countries	19 317	920	4.8	11	2.2
Africa	1 569	70	4.5	5	0.8
Latin America	8 147	313	3.8	28	8.6
Near East	2 637	168	6.4	24	4.4
Far East	6 964	369	5.3	6	1.4
Centrally planned economies	64 091	2 048	3.2	54	6.7
Asia	14 289	415	2.9	17	1.7
Eastern Europe and the U.S.S.R.	49 802	1 633	3.3	141	28.5
World	219 086	7 605	3.5	59	9.9

SOURCE: (81)

proportions were 18% for irrigation in the Near East and 6% for pesticides in the Asian centrally planned economies.

Data in Tables 17 and 18 indicate that seed production is the next largest consumer of commercial energy in agriculture; however, as its share is probably only a small fraction of 1%, it is not included in the present global and regional estimates.

Tables 22 and 23 also project the 1985/86 use of commercial energy for the four main input categories on the basis of recent trends. If these trends continue, the total world use of commercial energy in agriculture will almost double from 1972 to 1985; but this substantial

TABLE 22. – ESTIMATED AND PROJECTED USE OF COMMERCIAL ENERGY FOR INPUTS TO AGRICULTURAL PRODUCTION, 1972/73 AND 1985/86

	Fertilizer		Farm machinery		Irrigation		Pesticides		Total		Percent of world total	
	1972/73	1985/86	1972/73	1985/86	1972/73	1985/86	1972/73	1985/86	1972/73	1985/86	1972/73	1985/86
	.... 10^{15} joules											
Developed countries ...	1 635	2 800	2 851	3 355	57.0	66.7	93.6	107.4	4 637	6 329	61.0	47.0
North America	750	1 429	1 299	1 427	36.6	42.0	55.3	64.5	2 141	2 963	28.2	22.0
Western Europe	724	1 130	1 337	1 656	15.5	18.4	36.8	41.4	2 113	2 846	27.8	21.1
Oceania	35	69	100	121	1.3	1.7	0.7	0.7	137	192	1.8	1.4
Other developed countries	126	172	115	151	3.6	4.6	0.8	0.8	246	328	3.2	2.5
Developing countries ..	586	2 003	257	670	68.6	122.1	9.3	53.4	921	2 849	12.1	21.1
Africa	38	111	30	73	1.2	3.1	1.2	8.3	70	195	0.9	1.4
Latin America	153	468	148	349	6.1	13.7	5.3	13.8	313	845	4.1	6.3
Near East	86	351	50	167	30.8	54.7	1.4	8.3	168	581	2.2	4.3
Far East	309	1 073	29	81	30.5	50.6	1.4	23.0	370	1 228	4.9	9.1
Centrally planned economies	1 160	2 808	778	1 349	50.5	61.3	59.8	73.7	2 048	4 292	26.9	31.9
Asia	317	683	40	108	35.3	39.5	23.0	32.2	415	863	5.5	6.4
Eastern Europe and the U.S.S.R.	843	2 125	738	1 241	15.2	21.8	36.8	41.5	1 633	3 429	21.4	25.5
World	3 381	7 611	3 886	5 374	176.1	250.1	162.7	234.5	7 606	13 470	100	100
PERCENT OF TOTAL	44.5	56.5	51.1	39.9	2.3	1.9	2.1	1.7	100	100		

SOURCE: (81)

TABLE 23. – ESTIMATED AND PROJECTED TOTAL ENERGY REQUIREMENT AND PERCENT FOR EACH AGRICULTURAL INPUT, 1972/73 AND 1985/86

	Total		Fertilizer		Farm machinery		Irrigation		Pesticides	
	1972/73	1985/86	1972/73	1985/86	1972/73	1985/86	1972/73	1985/86	1972/73	1985/86
	10^{15} joules		*Percent*							
Developed countries	4 637	6 329	35.3	44.3	61.5	53.0	1.2	1.1	2.0	1.7
North America	2 141	2 963	35.0	48.2	60.7	48.2	1.7	1.4	2.6	2.2
Western Europe	2 113	2 846	34.2	39.7	63.2	58.2	0.7	0.6	1.7	1.5
Oceania	137	192	25.5	35.9	73.0	63.0	0.9	0.9	0.5	0.4
Other developed countries	246	328	51.2	52.3	46.8	45.9	1.5	1.4	0.4	0.3
Developing countries	921	2 849	63.6	70.3	27.9	23.5	7.5	4.3	1.0	1.9
Africa	70	195	54.3	56.9	42.9	37.4	1.7	1.6	1.7	4.3
Latin America	313	845	48.9	55.4	47.3	41.3	1.9	1.6	1.7	1.6
Near East	168	581	51.2	60.4	29.8	28.7	18.3	9.4	0.8	1.4
Far East	370	1 228	83.5	87.4	7.8	6.6	8.2	4.1	0.4	1.9
Centrally planned economies	2 048	4 292	56.6	65.4	38.0	31.4	2.5	1.5	2.9	1.7
Asia	415	863	76.4	79.1	9.6	12.5	8.5	4.6	5.6	3.7
Eastern Europe and the U.S.S.R.	1 633	3 429	51.6	62.0	45.2	36.2	0.9	0.6	2.3	1.2
World	7 606	13 470	44.5	56.5	51.0	39.9	2.3	1.9	2.1	1.7

SOURCE: (81)

increase would raise agriculture's share of the world total only slightly — from 3.5% to 4.1%.

In 1972/73 the developing countries (including the Asian centrally planned economies), with about two thirds of the world's population, accounted for only 17.6% of the total commercial energy used for agriculture.[1] This is very close to their share of the world's total commercial energy consumption. The 1985/86 projections indicate that energy use in agriculture would increase by 177% in the developing countries compared with only 56% in the developed countries. Thus the developing countries' share of total consumption would rise to 27.5% by 1985/86.

Commercial energy inputs for agriculture would increase most rapidly in the Near East and the Far East (more than 200%), and the increases would exceed 100% in each developing region. Among the other regions only eastern Europe and the U.S.S.R. would increase their share of the world total.

At the world level, fertilizers would move into first place by 1985/86, with a sharp rise in their share of the total commercial energy used in agriculture from 45% in 1972/73 to 56% in 1985/86. The share of farm machinery would fall equally sharply from 51% to 40%. The small irrigation and pesticides shares would decline slightly.

The share of commercial energy used for fertilizer production would rise in all regions (as much as 72% in the developing regions as a whole), remaining in first place in the developing regions and in second place in the developed regions. The share for the production and operation of farm machinery would fall, except in the Asian centrally planned economies. The proportion used for irrigation would decline in all regions, particularly the Near East. The share for pesticides would fall in the developed regions and rise in the developing regions, except in Latin America and the Asian centrally planned economies.

The purpose of the following review of the four main input categories is to see how commercial energy is used in producing and applying the input and whether opportunities for saving energy exist.

FERTILIZER

Chemical fertilizers may become the world's biggest user of the commercial agricultural energy. They already account for a high proportion in the developing countries owing to their predominant

---

[1] Agriculture in the developing countries utilizes large quantities of noncommercial energy inputs in the form of human labour and draught animals. Inclusion of these inputs might increase their total use of energy in agriculture by one fifth, but would only slightly raise their share of the world total.

use in present technologies to increase agricultural production by raising crop yields. Nevertheless, the proportion of total energy used for the fertilizers applied in the developing countries has been projected to rise only from 27% in 1972 to 35% by 1985 (Table 24). Chemical fertilizer production is even more unevenly distributed, only 17% being produced in the developing countries in 1972, when those countries relied on imports for 51% of their total supplies.

Nitrogen fertilizer is by far the most important chemical fertilizer in terms of the amount of plant nutrient used and even more so in terms of energy requirements. World consumption has been projected to rise from 36.2 million metric tons of nutrient in 1972 (28% in the developing countries) to about 84 million tons by 1985 (37% in the developing countries). Fertilizer is very energy intensive: a kilogram of nutrient requires about two kilograms of fossil fuel for its manufacture, packaging, transport, distribution and application (85).

In 1972 about 83% of all nitrogen fertilizer was produced in developed countries, and the developing countries produced only about 48% of their own consumption. New production capacity is more frequently being constructed close to abundant supplies of natural gas in some developing countries, particularly in the Near East.

In the many areas where the simplest and cheapest source of hydrogen is natural gas,[1] the price of nitrogen fertilizer corresponds to that of natural gas. Nitrogen fertilizer prices rose sharply, however, between 1972 and 1974 owing to cyclical industrial developments, resulting in inadequate production capacity in relation to demand, as well as to higher petroleum prices. By the end of 1975 production capacity had increased and demand had moderated, causing nitrogen fertilizer prices to drop to less than half their 1974 levels.

World consumption of phosphate fertilizer is projected to increase from 22.5 million metric tons of nutrient in 1972 (14% in the developing countries) to about 40 million metric tons in 1985 (28% in the developing countries). In 1972 about 87% of all phosphate fertilizer was produced in the developed countries, and the developing countries imported about 40% of their consumption.

Major deposits of phosphate ore exist in Morocco and the U.S.A. (Florida). The ore, which contains 12–15% P_2O_5, is concentrated to 35% (most of which reacts with sulphuric acid to produce superphosphate). Phosphate fertilizer is much less energy intensive than nitrogen. The energy required to mine, concentrate, process, package,

---

[1] Much of this reserve is now wasted. It has been estimated that 62% of the natural gas produced by OPEC members in 1972 was flared and that this quantity would be sufficient to produce five times the nitrogen fertilizer consumption of the developing countries in 1978 (259).

TABLE 24. – ESTIMATED AND PROJECTED COMMERCIAL ENERGY USE FOR FERTILIZER PRODUCTION, 1972/73 AND 1985/86

	For nitrogen[1]		For phosphate[2]		For potassium[3]		For all fertilizer		Percent of world total	
	1972/73	1985/86	1972/73	1985/86	1972/73	1985/86	1972/73	1985/86	1972/73	1985/86
	10^{15} *joules*									
Developed countries	1 352	2 400	188	263	95	137	1 635	2 800	48.4	36.8
North America	640	1 267	71	101	39	61	750	1 429	22.2	18.8
Western Europe	592	964	85	104	47	62	724	1 130	21.4	14.8
Oceania	16	34	17	30	2	5	35	69	1.1	0.9
Other developed countries	104	135	15	28	7	9	126	172	3.7	2.3
Developing countries	528	1 850	43	112	15	41	586	2 003	17.3	26.3
Africa	32	96	4	12	2	3	38	111	1.1	1.5
Latin America	128	398	18	52	7	18	153	468	4.5	6.1
Near East	80	331	6	20	—	—	86	351	2.6	4.6
Far East	288	1 025	15	28	6	20	309	1 073	9.1	14.1
Centrally planned economies	1 016	2 463	85	185	59	160	1 160	2 808	34.3	36.9
Asia	296	628	17	42	4	13	317	683	9.4	9.0
Eastern Europe and the U.S.S.R.	720	1 835	68	143	55	147	843	2 125	24.9	27.9
World	2 896	6 713	316	560	169	338	3 381	7 611	100	100

SOURCE: (81)

[1] Production of 1 kilogram (nutrient content) of nitrogen fertilizer requires 80×10^6 joules of energy. – [2] Production of 1 kilogram (nutrient content) of phosphate fertilizer requires 14×10^6 joules of energy. – [3] Production of 1 kilogram (nutrient content) of potassium fertilizer requires 9×10^6 joules of energy.

transport, distribute and apply one kilogram of nutrient is estimated at 0.33 kilogram of fossil fuel (145).

The total consumption of potassium, or potash, fertilizer is projected to rise from 18.8 million metric tons of nutrient in 1972 (9% in the developing countries) to about 38 million metric tons by 1985 (16% in the developing countries). It is usually manufactured from salts, such as potassium chloride, which occur in a nearly pure state in many parts of the world. Nevertheless, about 97% of the 1972/73 world total was produced in developed countries, and the developing countries imported more than 87% of their consumption. Since the ores are soft and often found near the surface, less energy is needed to mine them than for phosphate, although additional energy is usually required for enrichment. The total energy needed to mine, concentrate, package, transport, distribute and apply one kilogram of nutrient is estimated at 0.21 kilogram of fossil fuel (145).

Because chemical fertilizer accounts for such a large and increasing proportion of the total use of commercial energy in agriculture and is so important for raising crop yields with the existing technology, it is essential to examine how it can be used more efficiently. An important factor is the relative proportions of available supplies used in the developed and the developing countries. Generally, initial applications of nitrogen fertilizer result in substantial yield increases, but with heavier rates of application the yield response diminishes.

Thus in many developed countries additional fertilizer applications will bring smaller marginal returns than in most developing countries, where both yields and fertilizer use are low. In many developing countries considerable scope also exists for the more efficient use of fertilizer production capacity, including its energy efficiency.

There are numerous ways in which chemical fertilizers can be used more efficiently in both the developed and the developing countries, such as well-timed sowing and better water management. Improved methods of fertilizer application, including proper placement and timing to coincide with the nutrient demands of crops, will also increase efficiency. For example, experiments at the International Rice Research Institute in the Philippines show that when fertilizer is placed close to the root zone, application rates can be halved without any reduction in yields. The breeding of high-yielding varieties of cereals and other crops which are very responsive to fertilizer has contributed to more efficient use, as well as being a main factor in increasing the demand.

Research, now under way, to improve the conversion of solar energy by plants should make higher yields possible with less fertilizer. Other research offers hope that the biological fixation of atmospheric nitrogen through a symbiotic relationship with certain bacteria (at present found only in legumes) can be genetically transferred to cereal and other crops.

Chemical fertilizers are a comparatively recent phenomenon in agriculture. Until Chilean nitrate and Peruvian guano were introduced into European agriculture in the 1830s, and until superphosphate manufacture began in the 1840s, "artificial" fertilizers were limited to soot, bones, hoofs and horns, saltpetre and lime. The maintenance and restoration of soil fertility depended on such practices as shifting cultivation, fallowing, crop rotation, catch-cropping (especially with nitrogen-fixing legumes) and the recycling of crop and animal residues.

Because chemical fertilizers have become available at low prices and present the advantages of concentration, portability and adaptability to different soil conditions and crop requirements, the use of crop and animal residues to maintain soil fertility has steadily declined in the developed countries. In many developing countries, especially where there is no tradition of mixed crop and livestock farming, crop and animal residues have generally been used as noncommercial fuel rather than as fertilizer (82). In these countries agriculture modernization has sometimes proceeded directly to energy-intensive chemical fertilizers; however, a major exception is China, where despite a rapid increase in chemical fertilizer application the use of crop, animal and human residues is still substantial.

The growing population of the past century, particularly the last twenty-five years, could never have been fed (even at the present inadequate levels in many countries) without an increased reliance on chemical fertilizers. However, to offset the cost of commercial energy, it may be necessary to use more crop and animal residues as fertilizer when possible.

To overcome the difficulty of transporting and using as fertilizer even a small part of the available organic material, the developing countries must convert their farming systems to mixed livestock-crop husbandry, meet the heavy labour requirement, raise the low level of technology, create opportunities for profitably raising output, develop skills and alter cultural attitudes (59, 75, 76). In addition, many waste products are used as fuel to provide noncommercial energy. Therefore, the anaerobic fermentation process, now used to produce methane (biogas) for fuel, must be further improved so as to retain the plant nutrient value of the organic material.

FARM MACHINERY

In 1972, farm machinery accounted for the largest portion of commercial energy consumption in world agriculture. In the developing countries, however, it occupied second place after chemical fertilizer, and in the developed countries it will probably soon drop to that position.

Like chemical fertilizer, farm machinery is a newcomer among the inputs used for agricultural production. In primitive agriculture only human energy supplemented the basic conversion of solar energy performed by plants until animal draught power came into use in most areas. Human and animal energy still accounts for a large share of the total power used for agricultural irrigation in many parts of the world — perhaps about a quarter each in the developing countries (Table 25).

In Europe and North America in the eighteenth and nineteenth centuries, improvements in farm machinery, such as ploughs, seed-drills and reapers, were initially made for the use of animal traction. Animal, wind and water power was used to operate stationary machinery, such as threshers and grinders. Steam power was introduced for stationary farm machinery early in the nineteenth century and applied to field operations from about the middle of that century; but there was no effective replacement for animal traction in field operations until the internal-combustion engine (in the form of the agricultural tractor) was introduced at about the turn of the century.

Except at the most primitive level, both hand tool and animal draught technologies require some commercial energy for the manufacture of implements, but they consume a negligible amount of energy com-

TABLE 25. – DISTRIBUTION OF AGRICULTURAL POWER IN DEVELOPING REGIONS

Region	Total available power/hectare		Human	Animal	Me-chanical
	Kilowatts	*Horse-power*	.... *Percent*		
Africa	0.07	0.10	35	7	58
Asia (excluding China) ...	0.16	0.22	26	51	23
Latin America	0.19	0.25	9	20	71
PERCENT OF TOTAL			24	26	50

SOURCES: (81, 98)

pared with mechanical power technology, which uses large quantities of commercial energy in the manufacture and operation of farm machinery. Thus the rapid tractorization of agriculture in the developed countries over the last fifty years, combined with the more recent spread of mechanical power technology in the developing countries, has led to a substantial rise in commercial energy use for agriculture.

The number of four-wheel and crawler tractors used in agriculture is projected to rise from 16.1 million in 1972 (9% in the developing countries) to 20.6 million by 1985 (13% in the developing countries). Annual tractor production was 1.6 million units in 1972 (8% in the developing countries), and has been projected to increase to about 2.2 million in 1985. The total weight of farm machinery (including tillage, planting, cultivation, harvesting and other equipment) manufactured by the industry was estimated at 15 million metric tons in 1972 and is expected to rise to 20.5 million metric tons in 1985.

Table 26 gives estimates of the commercial energy (petroleum) required to operate farm machinery. The annual fuel consumption of a tractor depends on its size and the agricultural operations it performs. In North America, for example, the number of tractors has decreased since 1968 with the replacement of old tractors by fewer but more powerful new ones. Not only is the average tractor larger there than elsewhere, but tractors and other machinery are used to perform nearly all crop production operations. In the developing countries the average tractor is much smaller and used mainly for tillage and transport.

The commercial energy required annually to operate farm machinery is about twice that needed for its manufacture, although the proportion is somewhat lower in the developing countries. These countries' share of the total commercial energy devoted to the manufacture and operation of farm machinery was only 8% in 1972 and is projected to increase to no more than 15% by 1985.

The importance of power-operated farm machinery to world agriculture is threefold. First, it can effectively perform heavy operations such as deep ploughing and land clearance. Second, and more importantly, it can speedily perform crucial operations such as tillage and planting at the right time, so that yields are increased and losses due to pests or weather reduced (38). Speedy tillage and planting are essential in the semi-arid areas of the developing countries, where the total crop area depends on how much land can be prepared and planted in the brief period of uncertain rainfall, and in the subtropical and tropical areas, where multiple cropping is possible.

The third and most basic function of farm machinery is to replace human labour. This has been particularly important in the developed

TABLE 26. – COMMERCIAL ENERGY INPUTS FOR FARM MACHINERY
MANUFACTURE AND OPERATION

	Manufacture		Operation		Total		Percent of world total	
	1972/ 73	1985/ 86	1972/ 73	1985/ 86	1972/ 73	1985/ 86	1972/ 73	1985/ 86
	. . . .		10^{15} *joules*	. . . .				
Developed countries .	957	1 082	1 894	2 273	2 851	3 355	73.4	62.4
North America	429	473	870	954	1 299	1 427	33.4	26.6
Western Europe . . .	458	519	879	1 137	1 337	1 656	34.4	30.8
Oceania	35	42	65	79	100	121	2.6	2.2
Other developed countries	35	48	80	103	115	151	3.0	2.8
Developing countries .	99	254	158	416	257	670	6.6	12.5
Africa	11	27	19	46	30	73	0.8	1.4
Latin America	56	128	92	221	148	349	3.8	6.5
Near East	21	68	29	99	50	167	1.3	3.1
Far East	11	31	18	50	29	81	0.7	1.5
Centrally planned economies	248	441	530	908	778	1 349	20.0	25.1
Asia	15	41	25	67	40	108	1.0	2.0
Eastern Europe and the U.S.S.R.	233	400	505	841	738	1 241	19.0	23.1
World	1 304	1 777	2 582	3 597	3 886	5 374	100	100

SOURCE: (81)

countries, where agricultural labour is scarce and costly. Except in
limited areas, this function is unlikely to assume much importance
in the developing countries for many years.

On the basis of the United Nations medium population forecast,
as well as the fairly optimistic assumptions regarding the growth of
nonagricultural employment opportunities, the agricultural labour
force of the developing world will continue to increase until early
in the next century (77). In the meantime, however, rapid mechaniza-

tion of certain agricultural operations is urgently needed in the developing countries to reduce appalling human drudgery.

The developing countries should pursue more selective or appropriate mechanization policies than most of them have in the past. It is hoped that the energy crisis will lead the developing countries to adopt a more rational approach to the use of scarce capital and foreign exchange resources for agricultural mechanization (229, 230). However, it would be disastrous in many respects if the higher costs for farm machinery slowed down its use in the developing countries where it is essential for increased food and agricultural production.

In these circumstances the more efficient use of farm machinery is imperative. In the developed countries, recent attempts have been made to promote minimum tillage practices, and the energy requirements have been reduced by combining such operations as planting and fertilization. In the developing countries, as a result of faulty planning and management, often half of the tractors are idle while the other half are operated below capacity. Moreover, spare parts are often subject to taxation and delays. The design and manufacture of farm machinery better suited to the conditions of the developing countries would help alleviate this problem.

More effective use of human and animal draught power could reduce the need for mechanized power in many developing countries. As draught animals are fed on pasture and crops grown on the farm, little commercial energy is required for their use. However, both the shortage of land and the rising demand for livestock products tend to put this source of power in direct competition with human food supplies, at least in the more densely populated developing countries. Animals are also susceptible to diseases which limit their traction power and operating speed (in large areas of Africa they are excluded by trypanosomiasis carried by the tsetse fly). The local manufacture of improved equipment for use with draught animals could be of major importance in many areas.

Labour productivity could be greatly increased by the better design and utilization of hand tools, as well as by planning farm operations so as to avoid unnecessary seasonal peaks in labour requirements.

In the developing countries the effectiveness of all these power sources can also be increased by combined use. Whereas mechanized power is often best for tilling, animal power may be used for planting and secondary cultivation, and human labour for inter-row cultivation and harvesting. Mechanization in the developing countries should therefore complement rather than replace human and animal power.

IRRIGATION

There are two types of controlled irrigation: (1) large-scale gravity flow irrigation, where dams or water-diversion structures and channels are built to bring the water to fields, and (2) pump irrigation, where water is pumped from either groundwater or surface sources. As agricultural production is often only a secondary use of water from large dams — the primary use being electric power generation — the energy input for the construction of dams is excluded from the estimated inputs for agriculture. Because irrigation canals are constructed and maintained mostly with machines, the energy requirements for this purpose have been included under farm machinery. The following discussion is limited to pumps, engines, pipes and other irrigation materials, such as sprinkler equipment.

In 1972 the estimated quantity of irrigation equipment in use throughout the world amounted to about 2.5 million metric tons — almost half of it in the developing countries. By 1985 the quantity of such equipment is projected to increase to 3.6 million metric tons, about 2.2 million tons, or 60%, in the developing countries. The annual production of irrigation equipment supplied to agriculture is projected to increase from about 331 000 metric tons in 1972 to about 480 000 tons by 1985. It has been assumed that the energy required to produce this equipment is about the same as for farm machinery: 86.7×10^6 joules (2 kilograms of petroleum equivalent) per kilogram of equipment.

In addition to the energy required to manufacture irrigation equipment, energy (usually petroleum fuel) is needed for its operation (2). The fuel requirement per hectare of irrigated land varies with the depth of the pumped water, the type of irrigation system and the amount of water needed for the crop. Fuel requirements range from an estimated 160 kilograms per hectare in the developed countries to 200 kilograms per hectare in Africa and the Near East. The energy needed to operate irrigation equipment is about five times that required for its manufacture.

Irrigation is very important in the developing countries, especially in the Near East and the Far East (Table 27). Of the total commercial energy used for the manufacture and operation of irrigation equipment in 1972, the developing countries accounted for about 59%, and by 1985 this proportion is expected to reach 65%.

More efficient use of irrigation water, with savings in energy consumption, is necessary and feasible in many areas. Many existing irrigation schemes need considerable renovation. The interdependence of land development methods, irrigation practices and systems of

TABLE 27. – COMMERCIAL ENERGY INPUTS FOR IRRIGATION EQUIPMENT
MANUFACTURE AND OPERATION

	Manufacture		Operation		Total energy		Percent by region	
	1972/ 73	1985/ 86	1972/ 73	1985/ 86	1972/ 73	1985/ 86	1972/ 73	1985/ 86
	. . . . 10^{15} joules							
Developed countries ..	9.6	10.5	47.4	56.2	57.0	66.7	32.4	26.7
North America	6.0	6.4	30.6	35.6	36.6	42.0	20.8	16.8
Western Europe	2.8	3.0	12.7	15.4	15.5	18.4	8.8	7.4
Oceania	0.2	0.3	1.1	1.4	1.3	1.7	0.7	0.7
Other developed countries	0.6	0.8	3.0	3.8	3.6	4.6	2.1	1.8
Developing countries ..	11.9	21.3	56.7	100.8	68.6	122.1	39.0	48.8
Africa	0.2	0.6	1.0	2.5	1.2	3.1	0.7	1.2
Latin America	1.6	2.6	4.5	11.1	6.1	13.7	3.5	5.5
Near East	5.7	10.4	25.1	44.3	30.8	54.7	17.5	21.9
Far East	4.4	7.7	26.1	42.9	30.5	50.6	17.3	20.2
Centrally planned economies	7.2	9.8	43.3	51.5	50.5	61.3	28.6	24.5
Asia	4.2	5.6	31.1	33.9	35.3	39.5	20.0	15.8
Eastern Europe and the U.S.S.R.	3.0	4.2	12.2	17.6	15.2	21.8	8.6	8.7
World	28.7	41.6	147.4	208.5	176.1	250.1	100	100

SOURCE: (81)

cultivation and crop production is seldom fully appreciated. Both
increased efficiency in irrigation use and higher cropping intensities
can be achieved (1) by improving water distribution channels and
providing drainage, (2) by improving field layouts, (3) by grading
and levelling, and (4) by using better implements and water application methods in cropping. However, fragmented landholdings and
the difficulty of organizing the group action required to implement
new projects often impede irrigation improvements.

Extensive arid areas might be used for crop production only if irrigated with desalinated sea water. This would demand enormous quantities of energy, not only to desalinate the water, but also to transport it long distances from the coast. This development must therefore await the availability of less costly alternative energy sources.

PESTICIDES

World pesticide consumption (including herbicides, insecticides and fungicides) was estimated at 1.6 million metric tons in 1972, of which 319 000 tons (20%) were used in developing countries. By 1985, pesticide consumption is expected to be 2.3 million metric tons, of which 846 000 tons (36%) will be used in the developing countries.

The commercial energy required to produce a pesticide can be substantial. Raw materials for pesticides come mostly from the petrochemical industry. Manufacturing requires further energy inputs. A pesticide also contains a number of formulating agents and often a solvent, which also involve energy inputs. Packaging, transport, distribution and application require further energy inputs. The total energy required to supply a kilogram of pesticide has been estimated at about 101×10^6 joules, 2.4 kilograms of petroleum equivalent (144), which means pesticides are the most energy-intensive agricultural input (Table 28).

Concern about possible detrimental effects on the biosphere from chemical pesticide application has stimulated the search for ways of economizing on their use. Some alternatives to herbicide use are weed control through better tillage and mechanical or hand weeding. These methods are practicable in the developing countries, where labour is usually abundant and cheap compared with the cost of imported materials. Insecticide and fungicide use can also be reduced by developing new crop varieties with greater resistance to insects and diseases. Biological control by introducing an insect's natural enemies or sterile insects is finding wider application.

Nevertheless, a continuing increase in pesticide use is unavoidable, particularly in the developing countries, where crop losses (both pre- and post-harvest) due to inadequate pest control are large. The projections in Table 28 indicate substantial increases in pesticide use in each of the developing regions, except the Asian centrally planned economies. The declining share of world pesticide use projected for the developed countries reflects environmental concerns. The banning of numerous pesticides for environmental reasons has caused difficulties in the developing countries as suitable substitutes for them have not yet been evolved.

TABLE 28. – COMMERCIAL ENERGY INPUTS FOR PESTICIDE
MANUFACTURE AND APPLICATION

	Production [1] (10^15 joules)		Percent by region	
	1972/73	1985/86	1972/73	1985/86
Developed countries	93.6	107.4	57.5	45.8
North America	55.3	64.5	34.0	27.5
Western Europe	36.8	41.4	22.6	17.7
Oceania	0.7	0.7	0.4	0.3
Other developed countries	0.8	0.8	0.5	0.3
Developing countries	9.3	53.4	5.7	22.8
Africa	1.2	8.3	0.7	3.6
Latin America	5.3	13.8	3.2	5.9
Near East	1.4	8.3	0.9	3.5
Far East	1.4	23.0	0.9	9.8
Centrally planned economies	59.8	73.7	36.8	31.4
Asia	23.0	32.2	14.2	13.7
Eastern Europe and the U.S.S.R. ..	36.8	41.5	22.6	17.7
World	162.7	234.5	100	100

SOURCE: (81)

[1] Production of 1 kilogram of pesticide assumed to require 101.3×10^6 joules of energy.

Examples of energy use in agriculture

Two kinds of energy flow are apparent in agriculture. One is the flow of food energy from the photosynthetic process to the consumer's plate; the other is a conventional fuel energy flow that occurs in vegetal production and its transformation into food.

Some of the solar energy fixed through photosynthesis is lost, first in harvesting and in uncollected by-products and then during storage (e.g., of grain and forage); however, the major losses occur in the transformation of vegetal into animal products. Ruminant animals can convert vegetal material that is not digestible to humans into protein-rich food; but the number of vegetal calories needed to provide one calorie of meat or milk is high (Table 29), as most of the energy is used in animal metabolism or lost in manure.

Food processing and meal preparation transform animal and vegetal material into food products. Additional conventional fuel energy is needed at this step, and food losses are large. Finally, the caloric intake of humans (2 000 to 3 000 kilocalories per caput per day in most countries) is many times lower than the energy fixed by the crop originally.

The examples of energy use in agriculture presented below are based on data from three developed countries — the United States, the United Kingdom and France — and on an analysis of the situation in the developing countries, using Senegal as an example.

UNITED STATES

Agriculture in the U.S.A. currently produces more than enough food to feed its population. In 1974 the production from almost one third of the croplands was sold in export markets, thereby helping to meet global food needs and to balance U.S. imports of energy, materials and manufactured products. The $22 thousand million earned from agricultural exports in 1975–76 assisted in the purchase of $25 thousand million of oil imports (28).

Energy use in the U.S.A. is classified into four major categories: industrial, transportation, residential and commercial. According

TABLE 29. – ANIMAL EFFICIENCY: INPUT OF VEGETAL CALORIES NECESSARY TO PRODUCE ONE KILOCALORIE OF ANIMAL OUTPUT

Product	Vegetal input (kilocalories)
Milk	4.5
Beef, sheep meat	9.0
Hog	5.0
Eggs	4.5
Poultry	5.6

SOURCE: (35)

to a study by Booz, Allen and Hamilton, Inc., for the Federal Energy Administration (85), the food system uses 5.5% of the industrial energy, 4.4% of the transportation energy, 3.4% of the residential energy and 3.2% of the commercial energy (Fig. 24); thus the food system uses 16.5% of the nation's energy. Other studies report values of from 12% to 20%, depending on the exact boundaries set for the food system and on the extent to which indirect energy use (machinery, buildings, etc.) is charged to it. Production agriculture uses only 3–3.5% of the nation's energy (Fig. 25). Extensive data on energy use in U.S. agriculture in 1974 published by the Federal Energy Administration (86) provide figures by commodity, operation, state and month.

FIGURE 24
Major categories of energy use and consumption (shaded area) in the U.S. food system (1).

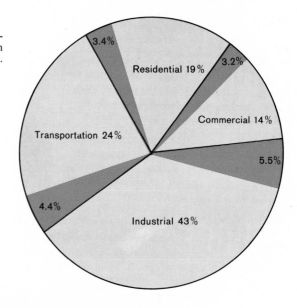

FIGURE 25
Percent of total U.S. energy consumption by industry (113).

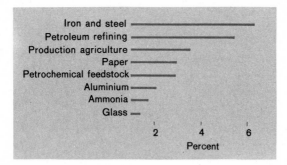

A comprehensive study by the Council for Science and Technology (232) not only answers many questions regarding current energy use in crop and livestock production, but also analyses alternative energy resources and their economic implications for the future.

Figure 26 shows the energy flow in the U.S. food chain for the production of one kilocalorie of human food (228). Of the potentially available plant energy, about 31 % is in nonfood crops such as cotton and tobacco, in crop residues such as corn stalks or straw and in exports. Thus, of every 16 kilocalories of potential plant energy in food, feed and fibre, only 11 actually enter the U.S. food chain, 1.2 kilocalories being used directly for human food and 9.8 for animal feed. After processing, transport, marketing and final food preparation, 62 % is consumed as vegetal products and 38 % as animal products. Of the original 16 kilocalories, only 1/16 is actually eaten by the U.S. population.

Figure 27 expands the above picture to show the total energy flow needed to feed the entire U.S. population for one year. For a population of 215 million people with a daily caloric intake of 3 000 kilocalories per caput, 236×10^{12} kilocalories per year are required.

Crop residues remaining in the field after harvest account for 25 % of the initial plant material. Of the energy left for domestic food and feed production, 94 % is used for animal feed. Only 15 % of this energy is from grain, however, and the rest comes from grazing land, hay and silage and by-products of food for human consumption. Of the energy used for animal feed, almost half is lost in manure.

Of the total fossil fuel input for food production in the U.S.A., only about one quarter is expended at the farm level. Three times as much fuel energy is used to process, transport, market and prepare food after it leaves the farm. Food preparation alone requires more energy than farm production; food processing uses about the same amount of energy as production; and the input of energy for marketing is only slightly less.

In 1970, 50 % of the energy used in the food system was liquid petroleum fuel — primarily Diesel fuel, gasoline and LP gas. Natural gas supplied 30 % and electricity 14 %. The remainder came primarily from residential fuel oil and coal (263).

Production agriculture

Fossil fuel energy (as distinguished from solar energy) is consumed in food production both off and on the farm. It is used off the farm in the manufacture of machinery, of nitrogen fertilizers (natural gas), of steel (coal) and of pesticides and plastics (petroleum). Fuel is used on the farm by tractors during tillage, planting, cultivation and harvest-

FIGURE 26. Energy flow in the U.S. food chain to produce one kilocalorie of human food energy (228).

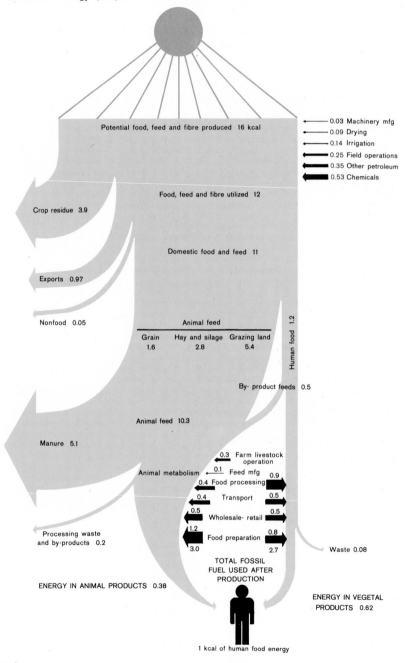

FIGURE 27. Energy flow in the U.S. food chain to produce a year's supply of food for 215 million people at 3 000 kilocalories per day (228).

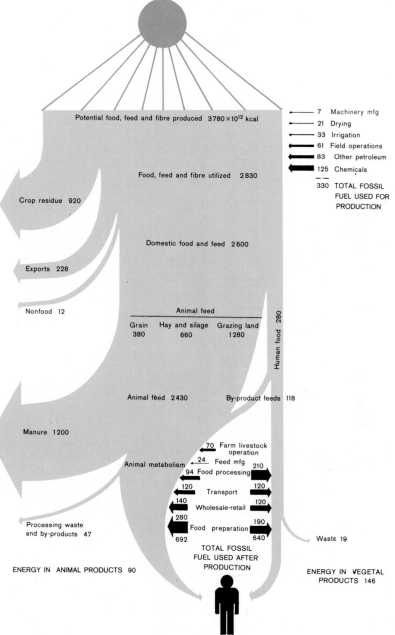

Potential food, feed and fibre produced 3780×10^{12} kcal

7   Machinery mfg
21   Drying
33   Irrigation
61   Field operations
83   Other petroleum
125   Chemicals

330 TOTAL FOSSIL
FUEL USED FOR
PRODUCTION

Food, feed and fibre utilized   2 830

Crop residue   920

Domestic food and feed   2 600

Exports   228

Nonfood   12

Animal feed

Grain   Hay and silage   Grazing land
380   660   1 280

Human food   280

Animal feed   2 430   By-product feeds   118

Manure   1 200

70 Farm livestock operation
24 Feed mfg
Animal metabolism   94 Food processing   210
120 Transport   120
140 Wholesale-retail   120
Processing waste and by-products   47
280 Food preparation   190
692   640

Waste 19

ENERGY IN ANIMAL PRODUCTS 90

TOTAL FOSSIL
FUEL USED AFTER
PRODUCTION

ENERGY IN VEGETAL
PRODUCTS   146

One year supply of human food energy for the U.S. population 236×10^{12} kcal

TABLE 30. – ON-THE-FARM ENERGY CONSUMPTION [1] IN THE U.S.A. FOR THE PRODUCTION OF FOOD, FEED AND FIBRE PRODUCTS

	Percent of total consumption [2]
Fuel	1.4
Electricity	0.4
Fertilizer	0.5
Agricultural steel	0.0
Farm machinery	0.5
Tractors	0.1
Irrigation	0.2
TOTAL	3.1

SOURCE: (263)

TABLE 31. – ENERGY CONSUMPTION IN THE U.S.A. FOR FOOD PROCESSING, PACKAGING AND TRANSPORT

	Percent of total consumption
Food processing	1.82
Machinery	0.01
Paper packaging	0.22
Glass containers	0.28
Steel and aluminium cans	0.72
Fuel for transport	1.46
Manufacture of trucks and trailers	0.44
TOTAL	4.95

SOURCE: (227)

Footnotes for Table 30: [1] Excludes domestic uses. Estimates from other sources indicate that the agricultural sector consumes 2–4% of the U.S. total. – [2] Based on 1970.

ing, as well as in pest control, frost protection equipment and irrigation pumps (Table 30 and Fig. 28).

Food processing

Food processing and related industries are collectively a major industrial energy user in the U.S.A. One study indicates that the food and fibre processing industry accounts for 7% of the total U.S. energy consumption (262). Another study shows that food processing, packaging and transport use only about 5% of the nation's energy (Table 31).

In 1973, natural gas supplied 48% of the energy used by fourteen leading food and kindred products industries (Table 32). Purchased electricity was second, supplying about 28%, followed by coal, the source of about 9%. The ratio of fuel types used by different industries varies considerably (e.g., 12% natural gas and 85% electricity in ice manufacturing, and 65% natural gas and 1% electricity in beet sugar processing). In addition, variations often occur among manufacturing plants in the same industry due to continuous versus batch processing,

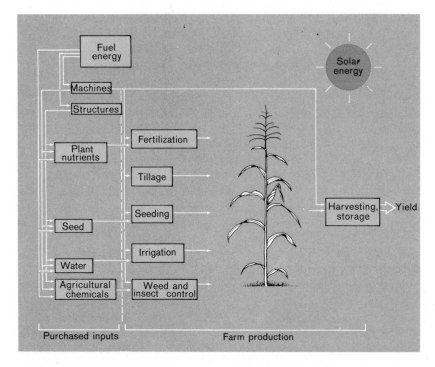

FIGURE 28. Energy flow in the production of row-crop products, U.S.A. (113).

plant and boiler design, drying operations, product concentration and plant location.

Commercial and home preparation

Finally, the purpose of the food system is fulfilled when the final food item is eaten. Energy is needed for all aspects of food preservation, preparation and waste disposal. A significant amount — nearly one and a half times the energy used in production agriculture — is required for commercial and home food preparation (Table 33).

UNITED KINGDOM

In 1968, agricultural production in the United Kingdom required 378.4×10^{15} joules (90×10^{12} kilocalories) of energy (Fig. 29), or 4.6% of that country's primary energy consumption (145). For this investment 130×10^{15} joules (31×10^{12} kilocalories) of human food were delivered, or enough to feed half the population (Table 34).

TABLE 32. – ENERGY USE BY FUEL TYPE OF FOURTEEN LEADING FOOD AND KINDRED PRODUCTS INDUSTRIES IN THE U.S.A., 1973

Industry	Natural gas	Purchased electricity	Petroleum products	Coal	Other
	. . . . *Percent*				
Meat packing	46	31	14	9	0
Prepared animal feeds	52	38	10	<1	0
Wet maize milling	43	14	7	36	0
Fluid milk	33	47	17	3	0
Beet sugar processing	65	1	5	25	4
Malt beverages	38	37	18	7	0
Bread and related products	34	28	38	0	0
Frozen fruits and vegetables	41	50	5	4	0
Soybean oil mills	47	28	9	16	0
Canned fruits and vegetables	66	16	15	3	0
Cane sugar refining	66	1	33	0	0
Sausage and other meat ..	46	38	15	1	0
Animal and marine fats and oils	65	17	17	1	0
Manufactured ice	12	85	3	0	0

SOURCE: (245)

TABLE 33. – COMMERCIAL AND HOME ENERGY INPUTS ASSOCIATED WITH REFRIGERATION AND COOKING IN THE U.S.A.

	Percent of total consumption
Commercial refrigeration and cooking ..	1.55
Refrigeration machinery	0.36
Home refrigeration and cooking	2.83
TOTAL	4.74

SOURCE: (227)

TABLE 34. – FOOD ENERGY OUTPUTS TO HUMAN CONSUMPTION IN THE
UNITED KINGDOM, 1968

	10^{15} *joules*
Wheat, maltsters and export	.06
milled flour	15.48
Barley, consumption and export	17.15
Oatmeal	1.06
Total cereals	**33.75**
Potatoes	12.22
Beet, refined sugar	14.81
Beetroot	.15
Carrots	.40
Parsnips	.07
Turnips and swedes	.07
Onions	.10
Total roots and onions	**27.82**
Brussels sprouts	.19
Cabbage	.48
Cauliflower	.21
Peas (shelled)	.68
Beans, broad	.03
runner and French	.04
Other vegetables: tomato, lettuce, celery, leek, cucumber, mushroom, etc.	.25
Total vegetables (except roots)	**1.88**
Fruit	**2.04**
Cattle, meat	9.99
offal	.36
Sheep and lamb, meat	2.83
offal	.14
Pigs, meat	12.71
offal	.16
Poultry	2.28
Total meat (dressed carcass weight)	**28.47**
Eggs, hen and duck	**4.61**
Milk, liquid	21.58
powder	1.71
condensed	4.00
cream	.55
butter	1.67
cheese	2.05
Total milk and products	**31.56**
General total	**130**

SOURCE: (145)

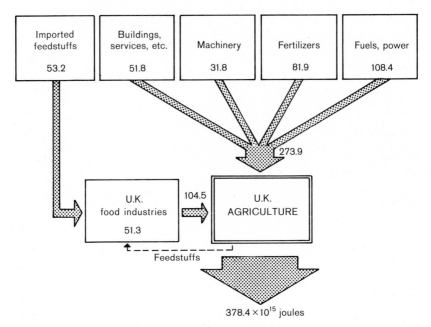

FIGURE 29. Energy inputs ($\times 10^{15}$ joules) to agriculture in the United Kingdom, 1968 (145).

Table 35 provides a detailed breakdown of the energy inputs to agriculture for primary food production. These inputs are only a minor fraction of the total energy needed to produce, process, transport, package and sell food in the United Kingdom and to import all the animal feed and human food that cannot be or are not grown domestically.

An estimate of these total energy inputs is shown in Figure 30. An overall total of nearly $1\,300 \times 10^{15}$ joules of energy for food production to the point of retail sale took 15.7% of the primary energy utilized by the United Kingdom. Agriculture plus domestic fisheries accounted for 31.6%; food processing, 36.6%; food and fish imports, approximately 21%; and food sales (including certain transport inputs), 10.7%.

In fact, the energy input for the total food chain is substantially greater. Once food is sold by a retailer, the buyer must carry it home, store it (perhaps in a refrigerator or home freezer) and cook it by gas or electricity. Other kitchen appliances also use energy, and so does the disposal of food remains and packaging, for example, and of sewage (to complete the food chain). These factors are not included in the estimated energy input. Adequate data for most of them —

TABLE 35. – ENERGY INPUTS TO AGRICULTURE IN THE UNITED KINGDOM, 1968

	10^{15} *joules*
Coal	5.62
Coke	3.31
Electricity	(49.59)
60% for nondomestic use	29.75
Petroleum: power units	
Diesel or gas oil	33.79
fuel oil	1.70
vaporizing oil	3.18
lubrication	0.91
50% motor spirit	6.01
Petroleum: heating, drying, etc.	24.15
(Total petroleum)	(69.74)
Total direct energy	**108.42**
Fertilizers: N	62.64
P	6.75
K	4.13
lime	8.40
Total fertilizers	**81.92**
Machinery, capital	21.29
noncapital	10.48
Total machinery	**31.77**
Chemicals	8.48
Buildings	22.77
Miscellaneous	4.28
Transport, services, etc.	16.28
Total other inputs	**51.81**
Feedstuffs: food industries	51.30
growing of imports	35.20
shipment of imports	18.00
Total feedstuffs	**104.50**
General total	**378**

SOURCE: (145)

FIGURE 30. Energy flow ($\times 10^{15}$ joules) for the total food system in the United Kingdom, 1968 (145).

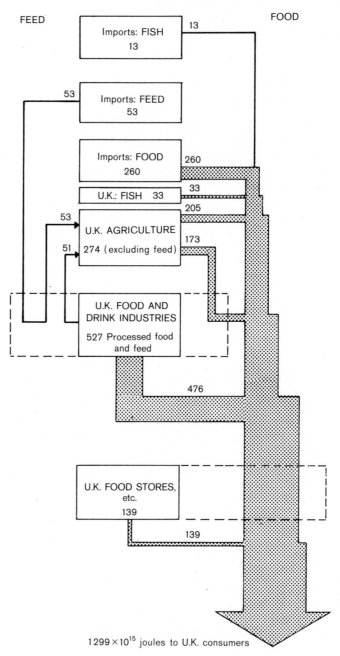

FEED   FOOD

Imports: FISH
13   13

53   Imports: FEED
53

Imports: FOOD   260
260

U.K.: FISH   33   33

53   205

U.K. AGRICULTURE

51   274 (excluding feed)   173

U.K. FOOD AND
DRINK INDUSTRIES

527 Processed food
and feed

476

U.K. FOOD STORES,
etc.

139

139

1299×10^{15} joules to U.K. consumers

particularly the transport of food shopping — do not exist, while cooking and refrigeration warm the home and often substantially reduce the energy consumed for space heating.

To illustrate the allocation of the energy flow in agriculture in the United Kingdom, Leach (145) presented a detailed breakdown of the energy inputs for a 1-kilogram loaf of white bread, sliced and wrapped (Fig. 31). Just under 20% of the total energy is consumed in growing the wheat, and all but 3% of the remainder is used in processing, packaging and transport.

In 1968, with an input of 1 300 × 10^{15} joules, the output of the United Kingdom food supply system was only about 261 × 10^{15} joules. This output is based on an average daily per caput intake of 2 560 kilocalories from food sources and 265 kilocalories from alcoholic drinks plus confectionery. Another 10% must be added for food and drink consumed outside the home (in canteens, restaurants, hotels, etc.). The overall total is therefore 3 107 kilocalories per person per day.

These figures give an overall energy ratio of 0.20, which means that five units of fuel are needed to supply each unit of dietary energy, excluding inputs beyond the grocery store. The figures also show

FIGURE 31. Percentage breakdown of the energy required to produce a 1-kilogram loaf of white bread and deliver it to a retail store in the United Kingdom. The total amount of energy required is 20.7 × 10^6 joules.

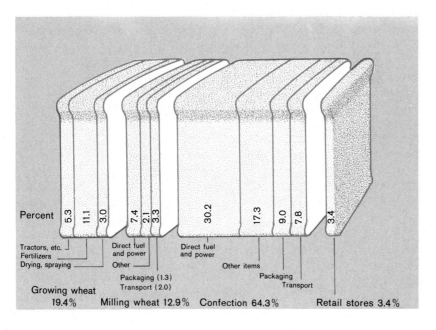

Percent 5.3 11.1 3.0 7.4 2.1 3.3 30.2 17.3 9.0 7.8 3.4

Tractors, etc.
Fertilizers
Drying, spraying

Direct fuel and power
Other

Direct fuel and power

Other items

Packaging (1.3)
Transport (2.0)

Packaging
Transport

Growing wheat 19.4%   Milling wheat 12.9%   Confection 64.3%   Retail stores 3.4%

that in 1968 the United Kingdom was almost exactly 50 % self-sufficient in dietary energy.

Clearly, this is not a viable energy input-output system for the developing countries. Copied on a global scale it would demand prodigious quantities of energy. If the present world population of 4 000 million each consumed 23.6×10^9 joules of fuels per year for eating, the total input would be $2\,247 \times 10^6$ metric tons oil equivalent, or 40 % of the global fuel consumption in 1972.

FRANCE

French agriculture is characterized by exportation, mainly to other European countries. As agricultural conditions and crops vary greatly from one part of the country to another, several types of agriculture are combined to provide an overall picture of the energy flow in the French food chain. Figure 32 draws from data compiled by Carillon (35). The data are similar to those in Figures 27 and 28 for the U.S.A., although some components are missing.

In France about two thirds of the harvested produce is used to feed animals. Although this proportion is less than in the U.S.A, and more than in the developing countries, it is high considering the losses inherent in transforming grain and forage into animal products.

For each kilocalorie of energy invested in production agriculture in France, 2.8 kilocalories of food and feed are produced. Of the 160×10^{12} kilocalories of input energy required for production, only 70×10^{12} kilocalories are direct fossil fuel.

The 300×10^{12} kilocalories used for animal feed are converted into only 25×10^{12} kilocalories of animal produce — a 1 to 12 conversion ratio.

The post-production (processing, transport, marketing and preparation) energy input is slightly greater than that of production. Of the total 175×10^{12} kilocalories of animal and vegetable matter leaving the production phase, 115×10^{12} kilocalories are in losses or in by-products, leaving only 60×10^{12} kilocalories in the food actually consumed. Thus about 1 kilocalorie of food energy is consumed for every 5 kilocalories of production and post-production energy inputs (0.18 calorie of food for each calorie of input energy).

Agriculture in France requires about half the energy input of U.S. agriculture and produces one sixth as much food, feed and fibre. Whereas in France 1 calorie of fossil fuel provides 2.8 calories of food, feed and fibre, in the U.S.A., 1 calorie of fuel produces 7 calories. But the higher meat component in the U.S. diet and the greater degree

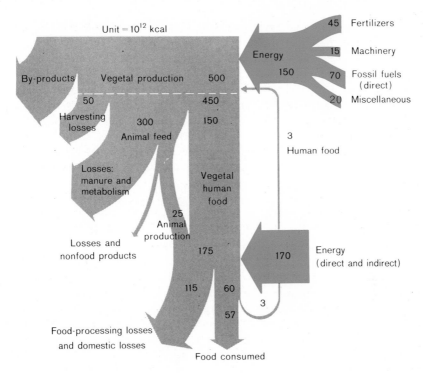

FIGURE 32. Agricultural energy flow in France (35).

of processing result in 0.14 calorie of food energy for each calorie of input as against 0.18 in France.

THE DEVELOPING COUNTRIES

As farmers in the developing countries need all the energy they can get to supplement their own manual capacity, energy planners should provide them with as much energy as they can afford and effectively use to increase productivity.

Energy flow data in the developing countries are often fragmentary. From data compiled on a number of prototypal villages in developing countries (see Table 37), Makhijani and Poole (151) pointed out that the total amount of energy, including animal and human labour, going into farming is surprising. Farmers in developing countries often use more energy per hectare than those in industrialized nations. For example, it takes more energy to feed the bullocks and mules that work the fields of these countries than to farm with the heavily mechanized methods of the developed countries.

As farms in the developed countries are often more productive per . unit of land, the energy use per ton of food is lower than in the developing countries, even when heavy irrigation and fertilizer inputs are considered. Of course, the amount of commercial energy used in agriculture in the developing countries is limited both by economics and by the lack of infrastructure and know-how.

Revelle (201) has pointed out that international statistics on energy are usually based on "commercial" energy consumption and hence seriously underestimate total energy consumption in poor countries. In India, for example, commercial energy use per caput is estimated at 150–190 kilograms coal equivalent, whereas the total energy use from all sources is nearly 490 kilograms.

In rural areas of poor countries the energy provided by human labour is five to ten times that obtained from commercial sources. Owing to the shortage of usable energy in such areas, there is a great need for both an increased supply and more efficient utilization.

Table 36 shows the energy intensity of rice cultivation in four major producing countries. The U.S.A. and Japan produce rice with one third the energy input per ton of India and about two thirds that of China, largely because of high yields in the U.S.A. and Japan due to the efficient use of water, fertilizer and machinery.

TABLE 36. – ENERGY INTENSITY OF RICE CULTIVATION IN
FOUR MAJOR PRODUCING COUNTRIES

Country	Farm machines and draught animals	Farm opera- tions	Irrigation and nitrogen fertilizer manufac- ture	Total input	Rice yield	Energy use per metric ton
	HP/ha	.... 10^6 kcal/ha			kg/ha	10^6 kcal
India	0.7	5	2	7	1 400	5
China	0.7	5	3	8	3 000	3
Japan	1.6	3	6	9	5 600	2
U.S.A.	1.5	2	6	8	5 100	2

SOURCE: (151)

Makhijani's and Poole's data for six prototypal villages in developing countries are presented in Table 37. Gross energy input per caput varied from 370×10^4 kilocalories per year in India to $1\,550 \times 10^4$ kilocalories in Mexico. The efficiency of converting gross energy input into useful work was about 5% in India and 25% in Mexico, as a result of which twenty times more useful energy was available per person in the typical Mexican village.

Studies show that in rural areas of the developing countries most of the energy is used for domestic cooking and heating (151). Wood and other vegetable matter, crop residues and cattle dung are the main domestic fuels. From an engineering standpoint this energy is used inefficiently and the yield of useful energy is small; from the economic standpoint of the villager, he may be making the best use of available resources.

The gross energy input for growing various crops in the six prototypal villages is shown in Figure 33. Tillage, irrigation and fertilizers are included, but not threshing, drying and transport. The gross energy requirements decrease where tractors are used in place of draught animals. Efficient irrigation and fertilizer practices increase energy inputs, but higher yields reduce the total energy requirement per ton of produce.

FIGURE 33. Energy intensity of farming in prototypal villages in developing countries (151).

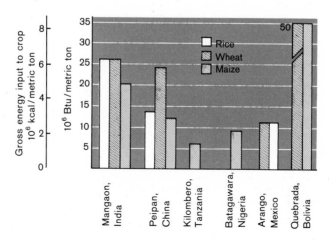

TABLE 37. – COMPARISON OF ENERGY USE PER YEAR IN SIX PROTOTYPAL VILLAGES IN DEVELOPING COUNTRIES (10⁴ kilocalories)

| Village | Domestic use per caput | | Agricultural use for farm work, irrigation, chemical fertilizers | | | | Use per caput for transport, crop processing and other activities | | Total per caput use | |
| | | | Per caput | | Per hectare | | | | | |
	Useful energy [1]	Energy input	Useful energy	Energy input	Useful energy	Energy input	Useful energy	Energy input	Useful energy	Energy input
Mangaon, India	5	101	13	194	40	645	3	86	20	370
Peipan, China	25	504	35	209	164	1 050	3	81	63	794
Kilombero, Tanzania ...	28	554	2	58	3	96	>1	18	30	630
Batagawara, Nigeria	19	378	4	60	10	184	1	23	23	466
Arango, Mexico	40	428	340	1 030	375	1 150	3	91	383	1 550
Quebrada, Bolivia	43	839	8	169	45	1 010	8	166	58	1 170

SOURCE: (151)

[1] Useful energy is defined as the amount developed at the plough or the shaft of the pump; it depends on the efficiencies of the various technologies of energy use.

Revelle (201) tabulated the use of energy in rural India (Table 38). Energy use per person in 1971 was 7.1×10^3 kilocalories per day, 3.3 times the energy in food consumed. More than 89% of this energy was from local sources, and less than 11% was from commercial sources.

TABLE 38. – ENERGY USE IN RURAL INDIA, 1971

Source of energy	Energy used (10^{14} kilocalories)					
	Agri-culture	Do-mestic	Lighting	Pottery, brick-making, metal-work	Trans-port and other	Total
Local						
Human labour	0.59	0.39		0.01	0.09	1.08
Bullock work	1.35				0.26	1.61
Firewood and charcoal .						4.60
Cattle dung		6.78		0.75		1.86
Crop residues						1.07
Total from local sources ..	1.94	7.17		0.76	0.35	10.22
Commercial						
Petroleum and natural gas						
Fertilizer	0.35					0.35
Fuel	0.08		0.42			0.50
Soft coke		0.14				0.14
Electricity						
Hydro [1]	0.03		0.01			0.04
Thermal [2]	0.12		0.05			0.17
Total from commercial sources	0.58	0.14	0.48			1.20
Total local and commercial	2.52	7.31	0.48	0.76	0.35	11.42
DAILY USE PER CAPUT (10^3) .	1.57	4.55	0.30	0.47	0.22	7.11

SOURCE: (201)

[1] Potential energy in water used to generate hydroelectric power. – [2] Energy in coal used to generate thermoelectric power.

Senegal

The energy flow in a typical traditional farming system (Fig. 34) and in an improved system using an animal toolbar (Fig. 35) has been charted from data published in 1972 for the Siné-Saloum area of Senegal (36). There, the annual rainfall ranges between 800 and 1 000 millimetres, and most of the indigenous crops (e.g., cotton,

Figure 34. Energy flow for an 11-hectare traditional farm in Senegal (36).

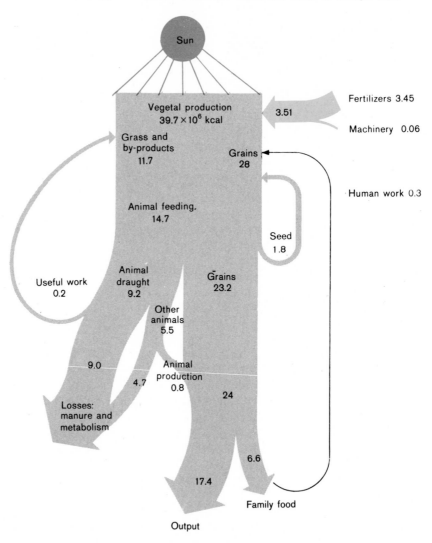

sorghum, millet, rice and maize) plus export crops (e.g., groundnuts) are grown there. The data include hours of work for each operation, fertilizer use and crop yields. The calculations for both the traditional and improved systems were based on the following conditions:

— 11 hectares of land: 3 hectares in groundnuts, 3 hectares in millet, 2 hectares in rice and 3 fallow hectares used for pasture;

FIGURE 35. Energy flow for an 11-hectare improved farm in Senegal (36).

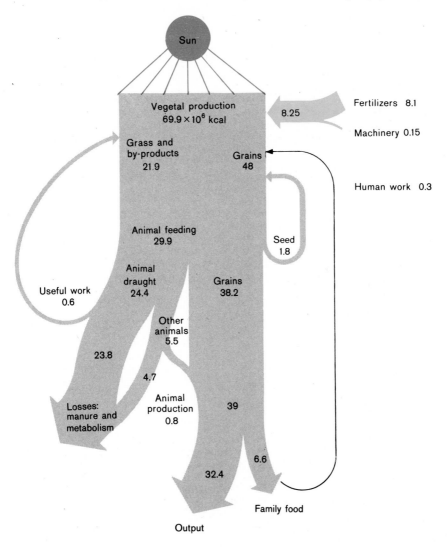

— animals, including chickens and goats, with a total weight of 250 kilograms;
— one family equivalent to six adults.

In the traditional system an animal weighing 300 kilograms is required to pull simple implements (maximum weight 30 kilograms), and in the improved system a pair of bullocks weighing 800 kilograms each is required to pull a multipurpose toolbar (total weight 75 kilograms). The same amount of seed per hectare was used under both systems (groundnuts, 55 kilograms; millet, 35 kilograms; rice, 100 kilograms), equivalent to 1.8×10^6 kilocalories.

Fertilizers: Inputs were calculated at 150 kilograms per hectare under the two systems (N-P-K formulations: 8-18-27 for groundnuts and rice; 14-7-7 for millet), plus 100 kilograms per hectare of urea for millet and rice under the improved system. The usual energy ratio for N-P-K production is equivalent to 3.65×10^6 kilocalories under the traditional system and 8.1×10^6 kilocalories under the improved system. No chemical for pest or weed control was used in either system.

Grain production: Typical yields (kilograms/hectare) were:

System	Groundnuts	Millet	Rice
Traditional	1 000	800	800
Improved	1 500	1 500	1 500

The production will thus have an energy equivalent of 28×10^6 kilocalories in the first case and 48×10^6 kilocalories in the second.

Draught animals: Harnessing time (hours/hectare) for draught animals was calculated as follows:

System	Groundnuts	Millet	Rice
Traditional	74	63	64
Improved	82	65	110

The traditional system requires 539 hours of animal work, assuming each animal's output is 0.48 kilowatt, equivalent to about 0.2×10^6 kilocalories of work per year. Under the improved system, 661 hours of work are provided by a pair of bullocks giving 1.1 kilowatts, equivalent to 0.6×10^6 kilocalories per year. The draught animals are maintained on agricultural by-products or pasture. Their extra energy need for the annual period of intensive work (4 months) is furnished by a grain supplement. Energy requirements for work are 2.5 times the nutrient value of maintenance feeding, equivalent in 10^6 kilocalories to the following per year:

System	Maintenance	Work	Total
Traditional	6.2	3	9.2
Improved	16.4	8	24.4

Machinery: Implements weighing 75 kilograms, with 10 years of life, will account for 0.15×10^6 kilocalories per year, assuming 20 000 kilocalories are needed to manufacture each kilogram of machinery; 30 kilograms of implements require 0.06×10^6 kilocalories.

Manpower: About 4 270 hours of work are needed in the traditional system and about 4 640 in the improved. Assuming a man can furnish continuous work of 0.07 kilowatt, the useful work will be equivalent in both cases to about 0.3×10^6 kilocalories.

Other animals: They are fed by-products or human food wastes. A bullock weighing 250 kilograms eats about 5.5×10^6 kilocalories per year. Animal production (meat, milk, etc.) requires 0.8×10^6 kilocalories per year (about 15% of the intake).

Humans: A family of six adults, each requiring 3 000 kilocalories of food every day, will eat about 6.6×10^6 kilocalories per year.

3. ENERGY CONVERSION

Introduction

Energy is constantly converted from one form to another by natural or human-controlled processes (Table 39) governed by the laws of thermodynamics. Basically, these laws describe the relationship between heat and mechanical energy, or work, and the conversion of one into the other. The first law of thermodynamics states that energy can be neither created nor destroyed, although it can change form. The second law explains that it is impossible to convert a given quantity of heat completely into work, as energy is always degraded in the conversion process, lessening its ability to do work. For example, to make a joule of high-grade electricity in a power station about 3 joules of heat must be released from chemical or nuclear fuel; thus 2 of the 3 joules are degraded to waste heat.

Some basic energy forms and conversion devices are listed in Table 40. Additional energy processes are given in Table 41 and Figure 36. The useful work (or energy) output compared with the energy input describes the efficiency of a conversion process, as shown in Figure 37 and Table 41.

This chapter deals specifically with various conversion principles and processes used in the food system. The description of each process is accompanied by a discussion of its efficiency and stage of development.

Solar energy

The sun radiates energy in the form of electromagnetic radiation. Each year an estimated 745×10^{15} kilowatt-hours of solar energy reach the outer atmosphere (9), equivalent to about 1.36 kilowatt-hours per square metre at 160 kilometres above the surface of the earth (14). These figures, however, represent only about one half of one thousand millionth of the total energy radiated by the sun (9).

TABLE 39. – E

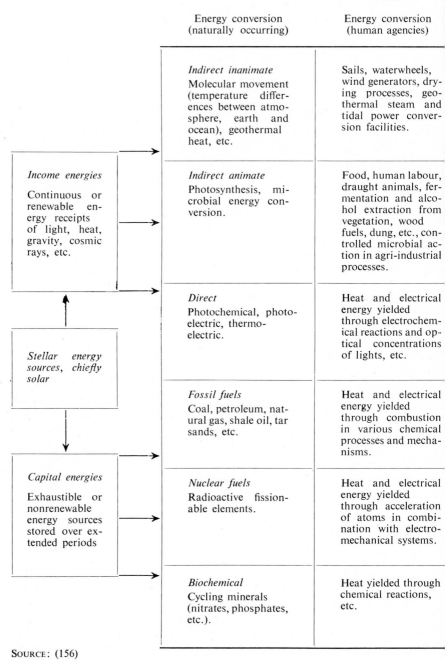

	Energy conversion (naturally occurring)	Energy conversion (human agencies)
Income energies Continuous or renewable energy receipts of light, heat, gravity, cosmic rays, etc.	*Indirect inanimate* Molecular movement (temperature differences between atmosphere, earth and ocean), geothermal heat, etc.	Sails, waterwheels, wind generators, drying processes, geothermal steam and tidal power conversion facilities.
	Indirect animate Photosynthesis, microbial energy conversion.	Food, human labour, draught animals, fermentation and alcohol extraction from vegetation, wood fuels, dung, etc., controlled microbial action in agri-industrial processes.
Stellar energy sources, chiefly solar	*Direct* Photochemical, photoelectric, thermoelectric.	Heat and electrical energy yielded through electrochemical reactions and optical concentrations of lights, etc.
	Fossil fuels Coal, petroleum, natural gas, shale oil, tar sands, etc.	Heat and electrical energy yielded through combustion in various chemical processes and mechanisms.
Capital energies Exhaustible or nonrenewable energy sources stored over extended periods	*Nuclear fuels* Radioactive fissionable elements.	Heat and electrical energy yielded through acceleration of atoms in combination with electromechanical systems.
	Biochemical Cycling minerals (nitrates, phosphates, etc.).	Heat yielded through chemical reactions, etc.

SOURCE: (156)

ERSION SYSTEMS

Advantages	Disadvantages	Human utilization
Winds, temperature differences, geothermal heat, etc. are continuous and have few inherent pollutants.	Difficult to control, periodic surges, geographic disproportion, etc.	→ Cooling food
Self-regenerating systems. Microbial action may be used in waste reclamation cycle.	Large volumes of heat energy dissipated in direct metabolic processes.	Preserving food → Cooking food Space heating
Direct energy transfer, may be stored biochemically or electrochemically, but on a relatively small scale.	Not yet capable of economic application on a wide scale or in large volume.	→ Space cooling Pumping water, air and gases
Stored, transported and controlled with ease and in large volumes.	Excess waste heat, gases and pollutants affect biogeochemical processes.	→ Motive power Lighting Communications
Independent of geography, minimal upkeep, by-product wastes may be used as other fuels.	Large investment in shielding and fuel refinement; disposal of radioactive waste is a key problem.	→ Industrial heating Conversion and forming of materials, etc.
May be stored, controlled and used as fertilizers in other food energy processes, etc.	Primarily a destructive energy use, as in explosives.	→

TABLE 40. – SOME BASIC FORMS OF ENERGY AND CONVERSION DEVICES

To \ From	Mechanical	Thermal	Acoustical	Chemical	Electrical	Light
Mechanical	Oar Sail Jack Bicycle	Steam engine	Barograph Ear	Muscle contraction Bomb Jet engine	Electric motor Piezoelectric crystal	Photoelectric door opener
Thermal	Friction Brake Heat pump	Radiator	Sound absorber	Food Fuel Match	Resistor Spark plug	Solar cooker Greenhouse effect
Acoustical	Bell Violin Wind-up phonograph	Flame tube	Megaphone	Explosion	Telephone receiver Loudspeaker Thunder	
Chemical	Impact detonation of nitroglycerine	Endothermic chemical reactions		Growth and metabolism	Electrolysis	Photosynthesis Photochemical reactions
Electrical	Dynamo Piezoelectric crystal	Thermopile	Induction microphone	Battery Fuel cell	Transformer	Solar cell
Light	Friction (sparks)	Thermo-luminescence	Rock music-light shows	Bioluminescence Candle	Light bulb Lightning	Fluorescence

SOURCE: (226)

TABLE 41. – TECHNICAL EFFICIENCY OF ENERGY CONVERSION

Devices	Electricity	Primary fuel
	.... *Percent*	
SPACE HEATING		
Gas furnace	—	75–85
Oil furnace	—	60–75
Coal furnace	—	55–70
Electric resistance heater	95	30–38
WATER HEATING		
Gas water heater	—	60–70
Oil water heater	—	50–55
Electric water heater	90–92	29–37
COOKING		
Gas stove	—	60–70
Electric stove	75	24–30
HOME APPLIANCES		
Gas clothes drier	—	45–50
Electric clothes drier	57	18–23
Gas refrigerator	—	30
Electric refrigerator	50	16–20
Gas air-conditioner	—	30
Electric air-conditioner	50	16–20
Small motor appliances	50–70	16–28
LIGHTING		
High-intensity lamp	33	11–13
Fluorescent lamp	23–28	7–11
Incandescent lamp	4	1.3–1.6
Mantle lamp (gas or kerosene)	—	0.5
TRANSPORTATION		
Diesel engine	—	38
Gas turbine	—	36
Automobile engine	—	23–25
Wankel rotary engine	—	18
Steam locomotive	—	8
MISCELLANEOUS		
Large electric motor	90–95	29–38
Large steam boiler	—	88
Storage battery (wet)	—	70–75
Fuel cell	—	60
Steam turbine	—	45
Gas laser	38–40	12–16
Steam power plant	—	32–40
Nuclear power plant	—	32–33
Solar cell	—	10

SOURCE: (226)

FIGURE 36. Energy conversion of various processes arranged on a logarithmic scale (243).

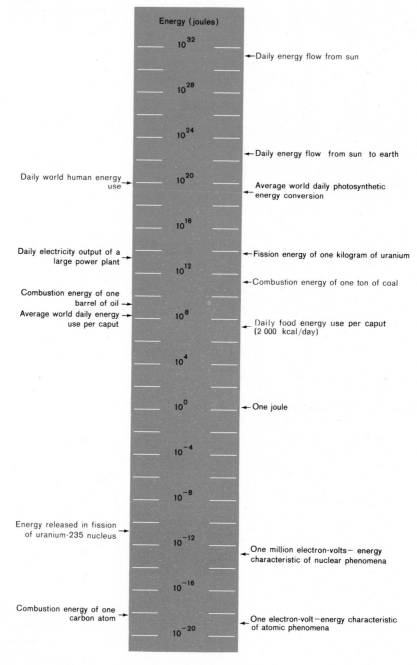

FIGURE 37. Efficiency of various energy converters. If the device involves a sequence of energy conversions, the total efficiency is given. For example, a steam-power plant involves the following sequence of conversions: chemical energy to thermal energy, thermal energy to mechanical energy, mechanical energy to electric energy. The overall efficiency of 40% is indicated by a bar across all three columns (235).

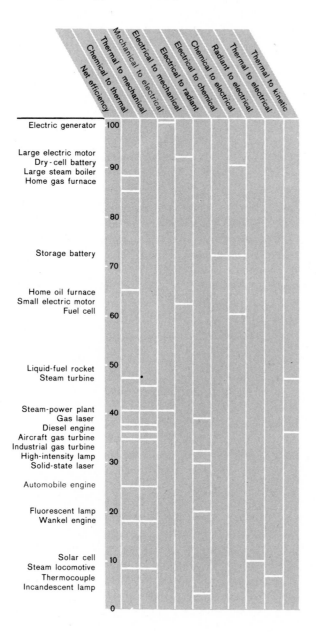

Of the solar energy that reaches the outer atmosphere only 6–8 kilowatt-hours per square metre per day can reach the earth's surface near the equator under optimum conditions (187, 226). Radiation intensity under cloudy conditions may be only 10–20% of that under clear sunshine. About 30% of the radiation reaching the atmosphere is reflected back into space. About 47% is absorbed as heat by the atmosphere, land and water, but much of this is reradiated back into the atmosphere.

Evaporation and precipitation use 23% of the atmospheric solar radiation. Wind and ocean circulation is also produced by solar energy, but uses less than 1% of the energy reaching the outer atmosphere. Surprisingly, less than 0.03% of this energy is used in photosynthesis.

Figure 38 depicts the power available from solar energy at different conversion efficiencies and for various land areas in the U.S.A., with the projected annual energy requirement to satisfy electric power

FIGURE 38. Availability of solar energy at different conversion efficiencies and for various land areas in the U.S.A. (173).

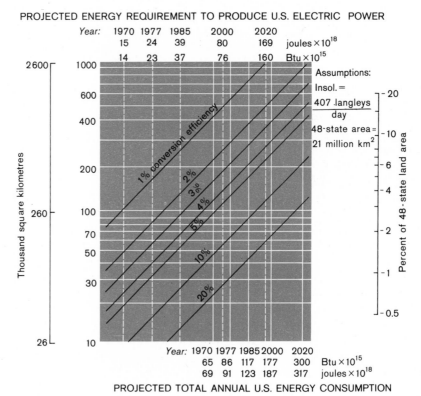

consumption and total energy consumption from 1970 to 2020. For example, if the solar energy falling on 26×10^4 square kilometres of land was converted into usable energy at 1% efficiency, it would provide the same amount of energy used by the U.S.A. to produce electric power in 1970.

PHOTOSYNTHESIS

Although the percentage of solar energy used by land and marine plants is small, the actual quantity is not negligible. Land and marine vegetation absorbs an estimated 4.2×10^{14} kilowatt-hours of solar radiation per year (9). World energy consumption is now about 6×10^{13} kilowatt-hours per year and is expected to reach 3×10^{14} kilowatt-hours by the year 2000 (9).

Photosynthesis is the mechanism by which plants utilize solar energy (Fig. 39). Energy from visible sunlight, in the form of electromagnetic radiation, is used by vegetation to take carbon dioxide (CO_2) and water (H_2O) from the environment and convert them into carbohydrates (CH_2O) and oxygen (O_2). This process is described by the chemical equation

$$CO_2 + 2H_2O + \text{light} \rightarrow (CH_2O) + H_2O + O_2.$$

Through photosynthesis the green leaves produce plant material and hence the world's food supply (48).

Respiration, another chemical process occurring in plants, is practically the reverse of photosynthesis and appears to reduce the efficiency of vegetation as an energy converter. During respiration the carbohydrates and oxygen combust and release carbon dioxide, water and energy, thereby consuming 20–40% of the gross photosynthesis (48, 97).

The efficiency of ordinary plants and trees is surprisingly low. A mature plant may convert sunlight into plant material at an efficiency rate of 1%; however, as a young plant cannot convert sunlight so readily, the lifetime efficiency is closer to 0.5%. Table 42 lists the typical photosynthetic yields of various ecosystems. Figure 40 shows the potential net photosynthesis expressed as plant material production for different climates.

Research is being conducted throughout the world in an attempt to improve the photosynthetic efficiency of plants. For example, laboratory experiments with algae have succeeded in increasing yields twentyfold over those obtained in ordinary agriculture (55). The prospect of improving photosynthetic yields for agricultural production is promising.

FIGURE 39
Solar energy (kilocalories) received annually by cropland in the U.S.A. Of every 28 000 kilocalories reaching the atmosphere, the earth's surface receives only about 16 000, of which only 7 500 are in wavelengths that can be utilized in photosynthesis. Because reflection, inefficiency, heat production and respiration use up most of this energy, in the end only 16 kilocalories of plant material are produced (174).

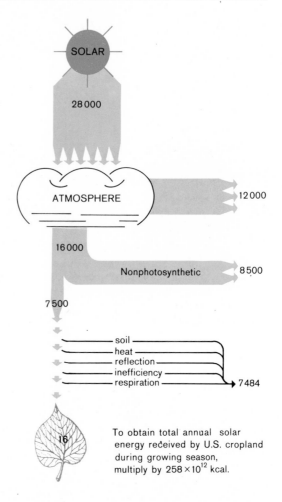

To obtain total annual solar energy received by U.S. cropland during growing season, multiply by 258×10^{12} kcal.

Although the use of solar energy as a power source is technically feasible, much research is still needed. Costs must be reduced, and simpler, more efficient applications are required. Until recently most industrialized societies showed little interest in solar energy conversion because of the ready availability of fossil fuels at low cost. Unfortunately, although many developing nations did not have easy access to fossil fuels, the development of alternative energy sources, including solar energy, was prohibited by a lack of technology.

Today almost the entire population of developing countries lives in an area bounded by the thirteenth parallel north and south of the equator (137). The surface of this area receives more solar energy than the rest of the world. Considering the lack of fossil fuels in

TABLE 42. – TYPICAL PHOTOSYNTHETIC YIELDS OF VARIOUS ECOSYSTEMS

Ecosystem	Grams/m²/day, dry organic matter, gross
Deserts	<0.5
Grasslands, deep lakes .... Mountain forests .... Some agriculture	0.5–3.0
Moist forests .... Shallow lakes .... Moist grasslands .... Moist agriculture	3–10
Some estuaries, springs, coral reefs .... Terrestrial communities on alluvial plains .... Intensive year-round agriculture (sugar cane)	10–25
Continental-shelf waters	0.5–3.0
Deep oceans	<1.0

SOURCE: (179)

FIGURE 40. Potential plant material production in various climates. During the summer months the production potential is lowest in the tropics, but over the year the tropics have one of the highest potentials because of the absence of cold winter months (97).

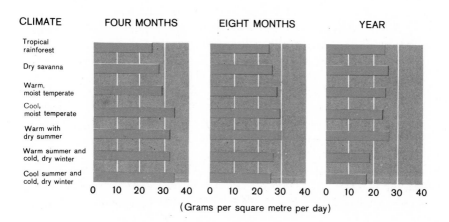

this area and the particular suitability of solar energy for small-scale applications, it appears that solar energy might make a major contribution to development in those countries.

CAPTURING SOLAR ENERGY

Plate collector

The simplest and most widely used method of obtaining solar energy is the plate collector, a sheet of blackened material set so that the sun's rays fall on it. Only black absorbs all wavelengths of visible light, as other colours reflect certain wavelengths. Typical collectors are rectangular boxes in which the black plate is placed with one or two layers of transparent material on the side exposed to the sun. The other sides are wood, and some insulation is placed in the bottom of the box. Air spaces are left between the black material, the transparent side, and the bottom. As the black surface is heated by the sun's rays the temperature of the adjacent air is raised. A medium such as water or air is sent through the collector for heating. This medium is then transferred to wherever heat is needed.

The designs for plate collectors shown in Figure 41 can be modified depending on use and location. Heat loss is an important consideration. As the difference in temperature between the black plate and the outside air increases, the amount of heat lost through conduction, convection and radiation also increases, thereby reducing the collector's efficiency. For warmer climates the examples in Figure 41 may be more elaborate than is necessary.

To intercept the greatest amount of solar energy, the collector should be tilted to the south if it is north of the equator and to the north if it is south of the equator. The tilt must be altered at least twice during the year to allow for differences in the angle between the horizon and the midday sun due to the changing of the seasons. For maximum collection during the winter the collector should be set at an angle equal to the latitude plus 15°, and during the summer at an angle equal to the latitude minus 15°.

To reduce construction costs, the collector can be built in a fixed position at an optimum tilt for the season of greatest use. Collectors are often set in a fixed vertical or horizontal position; this reduces efficiency, but how much depends on the location.

In a cold climate at 40° north of the equator a properly oriented and designed collector will be about 30% efficient. The efficiency may be cut in half for a collector in the same location but on a horizontal surface. In sunny weather a tilted plate collector can reach temperatures

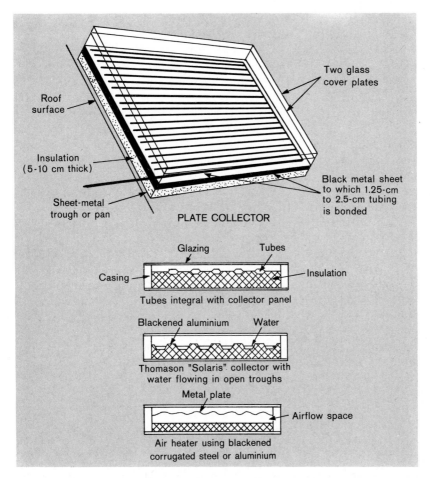

FIGURE 41. Plate collector. The black material is heated by the sun while a medium (air or water) is passed over the surface to pick up the heat and carry it to where it is needed. Below there are cross-sectional views of three designs: (A) water passing through the tubes gains heat from the black surface; (B) water flowing in open troughs is heated by the plate; (C) air is blown beneath the metal plate to heat it (14, 160).

of 70°–90°C. Higher temperatures are difficult to attain with flat plate collectors (55).

Focusing or concentrating collectors

Solar radiation reaches the earth's surface either as beam radiation in a straight line from the sun or as diffuse radiation scattered by clouds and dust. On clear days the radiation might be 90% beam, and on cloudy days it could be 100% diffuse (168). Focusing collectors can use only beam radiation, unlike flat plate collectors, which can use both beam and diffuse radiation. The focusing collector, however, can attain up to 3 500°C; even fairly crude ones can reach 500°C (9).

A focusing collector is usually curved to redirect the solar radiation that strikes it onto a smaller surface (Fig. 42). (This principle is exemplified by a magnifying glass held in the sun, which can burn a small hole through thin material.) The collector area can be made of any type of reflecting material. Curved mirrors are best, but metal, plastic or glass coated with aluminium are very effective, enabling the absorber to reach 500°C or more (55). The absorber should be as black as possible, capable of withstanding high temperatures and insulated along the top to prevent heat loss.

Most concentrators are parabolic in cross section, but other shapes are being studied. A simpler design looks like a long cylinder cut in half lengthwise. The inside surface acts as the reflecting area for the absorber tube running lengthwise through the centre (Fig. 43). A chain drive rotates the focusing collector to follow the path of the sun.

Because of the major difficulties associated with these collectors, few are being used. As they can use only direct sunlight, they are useless in cloudy weather. During the day the position of focusing collectors must be changed continually to follow the sun's path. This involves fairly complicated tracking equipment, which makes the cost prohibitive, especially in the developing countries.

FIGURE 42. Use of beam radiation by plate and focusing, or concentrating, collectors. In the focusing collector (*right*) the sun's rays strike the collector area and are redirected to an absorber at the focal point. The other two drawings show how beam (direct) and diffuse (scattered) radiation react with plate and focusing collectors. The plate collector's absorber accepts both direct and scattered rays, whereas the absorber of the focusing collector can utilize only direct rays (55, 158).

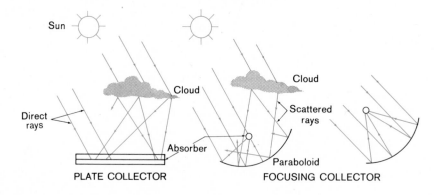

FIGURE 43
A parabolic-trough con-
centrator. Solar radiation
strikes the mirror area
and is reflected onto the
absorber tube in a more
concentrated state (14).

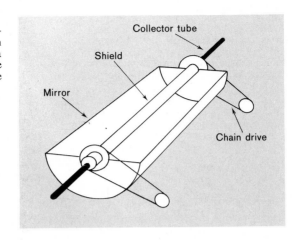

However, in areas where money is scarce, labour is often abundant
and relatively inexpensive. It is possible to track the sun by hand
operation provided the collector is shifted 15° every hour (55). Another
possibility is a mechanical tracking device controlled by a pendulum
like that of the grandfather clock. Hand power would be needed only
to lift a 20-kilogram sack of sand one metre each day to provide the
power necessary to rotate the collector (55). Some focusing collectors
are fixed in position and require no tracking, but need to be shifted
with the seasons. At present these are inefficient and require more
research.

Another problem is the precision needed in construction. Moreover, wind, rain, dust and other inclement conditions must be properly
guarded against. Because these factors increase costs considerably,
focusing collectors are uneconomic today.

Solar ponds (*collection and storage*)

The collectors described above would be useful in small-scale operations, whereas a solar pond could provide energy for a small community.

The principle of solar ponds is simple, but it was not until the 1950s
that a few observers began to note and relate the underlying natural
phenomena. In an ordinary shallow pond solar heat passes through
the water and raises the temperature of the bottom layers, thereby
lowering the water density and causing it to be replaced by heavier
cold water from above. As a result, convection currents are set up
which disperse the heat throughout the pond (Fig. 44). Consequently,
no portion of the pond reaches a significantly higher temperature and
heat loss is substantial across the water-air interface (239).

FIGURE 44. Solar heating of an ordinary shallow pond. Sunlight heats the bottom layers, reducing their density. This creates convection currents which dissipate the heat throughout the pond. The temperature difference between the surface and bottom of the pond is only 2°C (239).

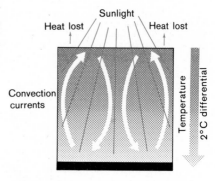

As salt water can form relatively stratified layers of various densities, a solar pond is made by forming density layers of salt water in a shallow basin about one metre deep. The bottom of the pond is lined with a black material to absorb the sun's energy (Fig. 45). Thus the bottom, heavier layer of salt water is heated and the gradient in the pond becomes strong enough to discourage convection currents. As the heat cannot be dissipated throughout the pond without these currents, the bottom layer can reach a considerably higher temperature.

Another useful property of salt water gradients is that any layer can be drawn off without disturbing the others. To obtain energy from the pond, the heated bottom layer can be pumped out, sent through a heat exchanger and pumped back to its original position in the pond. It takes about two years for a pond to reach its full value, but substantial energy can be obtained even several months after construction. In the winter only about a fourth as much heat can be withdrawn as in the summer.

Although the first solar pond was built in 1960, only a handful of people have conducted research on them, including groups in Israel,

FIGURE 45. Solar pond. Sunlight is drawn to the black absorbing material lining the bottom, thereby heating the bottom layer of salt water. Because the salt gradient throughout the pond prevents convection currents from forming, the temperature difference between the surface and the bottom of the pond is 70°C (239).

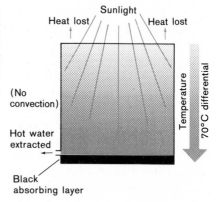

Chile, Australia, the U.S.A. and the U.S.S.R. (217). The Israeli experiments indicate an efficiency rating over a year of 25% (the ratio of heat delivered to total solar energy incident on the pond).

Solar ponds show promise, but there are technical problems that need to be resolved. The salt gradient still exhibits a degree of fragility. Methods of diminishing the influences of wind, storms, dust, rain and evaporation must be found. Furthermore, even though solar ponds are based on simple principles, they must be built by technicians. Yet, the potential of solar ponds is great, especially in developing areas.

The heat extracted from a solar pond can be used, for instance, to operate a low-temperature turbine. Calculations show that with a 25% efficient one-square-kilometre solar pond, a turbine of only 8% efficiency could operate a plant of about 5 000-kilowatt capacity during the summer and of 20–25% that capacity during the winter. These calculations assume a latitude of 32° N; there would not be such a sharp reduction in winter capacity closer to the equator. Solar ponds might also be used for desalination and any other applications requiring large amounts of cheap low-temperature heat.

Solar ponds differ from the other methods of collecting solar energy inasmuch as they not only capture solar radiation but also store it without external devices.

Heat storage

A major difficulty with all solar energy systems is that solar radiation is extremely unreliable. As many applications require a fairly constant energy source, it is necessary to either store heat collected in favourable weather or rely on a conventional source, such as a fossil fuel, as a back-up system.

The storage of solar heat is a problem. Because solar radiation is a dilute energy source, the more economically feasible devices for capturing this energy can only attain temperatures of less than 100°C. Heat storage systems are likewise limited to this temperature range and consequently require large areas of space. The storage area is generally filled with inexpensive materials, such as water, rocks or soil, which tend to retain heat for fairly long periods. As the heated water or air from the collector passes through the storage area, it raises the temperature of the storage material.

Water has the highest heat capacity of any ordinary material. A litre of water requires a kilocalorie of heat to raise its temperature 1°C. Thus a litre of water as it cools returns a kilocalorie for each degree of cooling. Rocks store about a third as much heat as water (55). In most areas water is inexpensive, but its use as a storage medium requires large, expensive tanks to hold the tons of water required. In many cases a circulating pump is also needed. A rock bed can be made by simply filling a container with gravel, crushed rock, or brick sifted to uniform size. Enough air space must be left to prevent an excessive resistance to air flow.

Large water tanks and rock beds can be placed underground because heat conductivity through the earth is low, although moist soil tends to cause more heat loss than dry soil. For small-scale storage systems the heat loss to the surrounding soil from underground tanks is too large.

Other mediums for heat storage are chemicals and metals. The use of chemicals tends to be expensive and complicated, and metals are also costly. For these reasons they are seldom used except in more elaborate systems.

SPECIFIC APPLICATIONS

Water heating

Water heating is one of the few commercially successful solar applications. An estimated several million solar water heaters are currently in use (9), over two million units in Japan alone (138). Like other solar energy devices, the units vary in efficiency, capacity and costs, but all water heaters use flat plate collectors to capture solar heat.

One of the simplest designs, widely used in rural Japan, holds 180 litres and attains temperatures of 55°C in summer and 27°–35°C in winter (55). The heater consists of a wooden tray lined with black polyvinyl plastic and covered by a hinged glass plate (Fig. 46). It is 0.9×2 metres and holds 10 centimetres of water.

Another heater consists of black plastic in the shape of a pillow. It is filled with water, set in the sun and emptied when hot water is needed. A standard size is $90 \times 180 \times 12$ centimetres and holds 190 litres of water. This heater will last about two years (55).

Figure 47 shows a more complex solar water-heating system than that in Figure 46. The type of storage tanks, insulation, materials and design chosen depends on need and cost. The layer of water must be shallow, preferably 5–10 centimetres. In sunny, warm weather the water temperature rises to about 50°C in three to four hours. Efficiency ratings vary depending on design but are good even for simple constructions. Efficiencies of 50–70% are not unusual (164).

Water distillation

Since the first large solar distillation plant was built in Chile in 1872, the design has little changed (240).

FIGURE 46
Solar water-heating tray. The wooden tray lined with black plastic is filled with water, and the glass cover is lowered. Solar radiation becomes trapped inside, heating the water to about 50°C within three or four hours in summer. When hot water is needed, it is drawn off through the outlet tube (55).

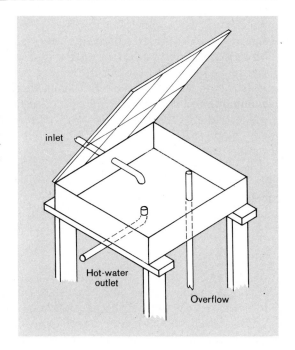

inlet

Hot-water
outlet

Overflow

FIGURE 47. Solar water heater with plate collectors to capture the sun's energy. Water is passed through the collectors for heating and then flows into a storage tank. The water transfer is caused by the thermosiphon effect created by the differences in water temperature and density (164).

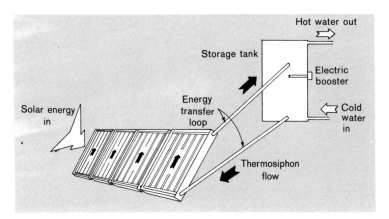

Hot water out

Storage tank

Electric
booster

Solar energy
in

Energy
transfer
loop

Cold
water
in

Thermosiphon
flow

Solar stills are a simple means of obtaining drinkable water from saline or brackish water. Basically, they consist of a blackened tray with a glass or plastic cover. Insulation is sometimes needed underneath the tray. As the impure water in the tray is heated by solar radiation it evaporates and condenses on the cooler transparent cover, eventually running down into collecting channels on the lower edges (Fig. 48).

Water storage facilities will probably be needed to ensure an adequate supply during the winter. By slightly modifying the solar still it can also be used as a rain trap.

The efficiency of most stills is about 35% (9). In sunny areas at 30° to 40° latitude an annual yield of about 1 cubic metre per square metre of collector can be expected (240). On a village scale, solar distillers are competitive with other means of obtaining fresh water in both initial construction and maintenance costs (240).

Water pumping

Water and fertilizer are required for adequate food production. Farmers in the developing countries are so poor that they often cannot afford costly fuels and fertilizers; thus food supplies are threatened.

In Southeast Asia a single crop of rice is grown with gravitational water during the rainy season. It would be feasible, however, to grow

FIGURE 48. Solar distillation. In the solar still, salt water is placed in a concrete container lined with a black material. As the water heats, it vaporizes and the impurities are left behind as residue. The vapour then condenses on the colder cover and runs down into the collecting troughs (9).

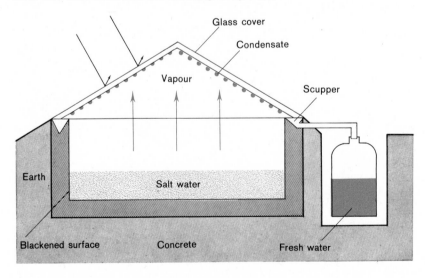

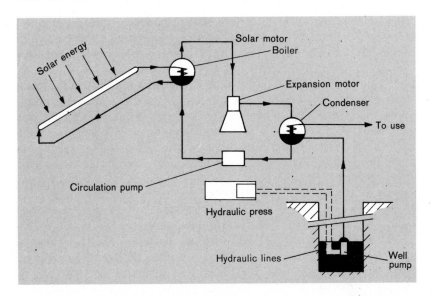

FIGURE 49. Thermodynamic solar water-pumping system. Heated water from the solar collector passes through a heat exchanger. The caloric content of the water is transferred to a fluid that circulates through the engine. The expansion of the fluid due to its temperature gain provides mechanical energy to power the engine, which drives a hydraulic ram that lifts the water (100).

up to three crops per year if water were made available during the dry season, thereby doubling the annual rice output without the use of fertilizers.

In many regions, including Thailand, main water and irrigation canals already exist, but secondary distribution systems are poorly developed. Therefore, 70% of the farmers need to pump water to secure a second or third crop. These areas require irrigation pumps with a 2- to 3-metre lift capacity. Water pumps with a higher lift capacity are needed in drier regions of the world and in upland areas of Southeast Asia (51).

In many arid regions water lies 30–40 metres below ground level (9). Low-power water pumping installations could usually supply enough water to sustain the small human and animal populations living in those areas (9).

Often, however, capital expenditures and increased management are needed to obtain more efficient irrigation or pumping. As energy costs rise, the costs of realizing and maintaining high efficiencies also increase (142).

Figure 49 depicts the basic components of a French solar water-pumping installation (100). Water is circulated through flat plate collectors and then sent to a heat exchanger which transfers the heat content of the water to the fluid circulating in the engine circuit. The engine powers the pumping system, which generally consists of a hydraulic press and a well pump.

Researchers in India are experimenting with solar lift pumps that have no other moving parts but check valves. No auxiliary power source and technical skill are required to run the pumps, which makes them most suitable for rural lift irrigation (96). In principle, a liquid confined in a closed tank is vaporized and moves to a higher elevation depending on the saturation pressure. By condensing the vapour below its normal boiling point, the liquid can be drawn into the tank from a source close to its level (199).

The Societé française d'études thermiques et d'énergie solaire has sponsored at least 35 solar water-pumping installations in Africa and Latin America (99). Each one-kilowatt station has the capacity to raise 30 cubic metres of water to a height of 20 metres per day. This will supply 1 500 people with 20 litres of water each per day or 750 cattle with 40 litres per head per day (99). A one-kilowatt solar water-pumping station requires 70 square metres of collector area and a two-cylinder 1 000-cc expansion engine. The amount of water pumped is based on a mean solar incident radiation of 700 kilocalories per hour per square metre, a 20°C air temperature and 5–7 hours a day operating time.

The initial cost of solar-powered pumps is high and cannot compete with Diesel or gasoline engines where fuel supply and maintenance are no problem. In other areas, however, the prospect of using solar energy to lift water is promising. Each particular case is different and should be considered independently. Among the advantages of solar power installations are minimum maintenance, absence of pollution, simplicity and fixed costs (162).

Diesel engines are more efficient than solar heat engines — which have an efficiency of 10% or less — and are initially less expensive. The cost of a Diesel engine and pump system for a one-kilowatt station ranges from $0.40 to $0.45 per cubic metre of water pumped. A solar-operated pump averages, instead, from $0.50 to $0.60 per cubic metre, but the cost is expected to drop in the near future to $0.10 to $0.14 per cubic metre (99).

Experiments using solar energy to lift large volumes of water for irrigation are under way in Mexico (3). A 30-kilowatt station was built for a turbine rather than an expansion engine. The turbine drives an alternator to produce electric power, which is used to pump water (99). In operation since September 1975, it is a mammoth installation and needs further study to make it economical.

Crop drying

Generations of farmers have effectively used the sun and the wind to dry their crops by spreading them out in thin layers on the ground. Solar radiation heats the surrounding air and changes the equilibrium vapour pressure, causing moisture to diffuse from the plant (158). A moderate wind accelerates the process by forcing the heated air through the crop and carrying the moisture-laden air away. Modern technology still uses heat and air flow for crop drying (30) but provides higher temperatures and better permeation (30, 32).

Two different types of crops need drying. Grains at harvest usually have a moisture content of 20–30%, which needs to be reduced to 12% before storage. The other category, leafy crops and fruit, has a moisture content of over 50% at harvest (158). As their moisture content must be reduced sharply before storage, during the extended drying process these crops are exposed to dirt contamination, to rain, to overheating and to fungal and bacterial growth (55). Therefore, further research is needed before solar drying can become practical for this category of crops, but it is practical now for grains.

Low-temperature solar grain drying is accomplished in many ways (218). The simplest method is to spread the grain in thin layers on the ground. This method can be improved by placing the grain on a dark surface, which will absorb incident radiation more readily than a light surface.

The International Rice Research Institute in the Philippines conducted solar drying trials during the dry season in 1974 with five different types of surfaces placed on an earth floor: concrete or cement, clear polyethylene sheeting, woven matting, synthetic jute sacking, and asphalt. A mechanical drier was used as a control. Two types of rice, long and short grain, were placed to a depth of 3 centimetres on the surfaces. The initial moisture content of 23–24% was to be reduced to 14% at most. The results are given in Table 43.

The rice dried fastest on asphalt, followed by cement or concrete, woven matting, synthetic jute sacking and clear polyethylene sheeting; however, the task was completed in 4–5 hours regardless of the type of surface. Mechanical drying, by reducing sun-checking in the grain, resulted in less breakage during milling. Woven mats and jute sacks, being perforated, permit greater aeration (increasing the rate of moisture reduction) than a solid surface like polyethylene. The results of this experiment, though tentative, give a good indication of the available possibilities.

In many areas where grain is dried only during a few months of the year the use of elaborate solar driers may not be economical, but some

TABLE 43. – RATE OF MOISTURE REDUCTION FOR SHORT GRAIN AND LONG GRAIN RICE
WITH ALTERNATIVE DRYING METHODS

Drying method	Short grain					Long grain				
	Drying time (hours)					Drying time (hours)				
	1	2	3	4	5	1	2	3	4	5
	.... *Percent moisture content*									
Concrete/cement	19.1	18.0	14.8	13.2	13.0	19.2	17.3	14.5	13.0	12.9
Polyethylene sheeting (clear)	19.0	16.8	15.7	14.1	13.8	18.5	17.0	16.0	14.4	13.8
Woven matting	17.0	15.5	14.6	13.5	13.6	18.3	15.8	15.5	14.3	13.9
Synthetic jute sacking ...	17.8	16.0	15.5	14.1	14.1	17.7	15.4	15.0	14.3	14.0
Asphalt	17.9	15.7	14.4	13.3	13.3	18.0	16.8	15.1	13.8	13.7
Mechanical drying (control)	17.7	16.1	12.3	—	—	16.0	14.7	13.6	—	—

SOURCE: (58)

relatively inexpensive, simple alternatives are available. The cost of special solar drying equipment is often justified, however, by the improved quality of the food product, as the shorter drying time reduces both the loss of vitamin C through oxidation and the production of materials with objectionable flavours (55). Draping a clear plastic cover over grain that has been spread on rocks to dry in the sun not only reduces the drying time but also keeps out dirt and insects, which cause a severe loss of produce (240).

In a few of the developed countries research is being conducted on solar drying techniques. One design that is gaining acceptance uses a long, black polyethylene tube as a solar collector (Fig. 50). The plastic tube is stretched out in the sun, and a fan forces air through it. The black material absorbs solar radiation and, in turn, heats the air. This warm air is then circulated through a bin containing the grain. Covering the black plastic layer with clear plastic reduces heat loss. Calculations show that in good weather at 45° N an energy equivalent of 109 kilowatt-hours per day can be obtained from a collector tube 45 metres long and 1 metre in diameter (158). This drier appears to be feasible and beneficial for drying both maize and soybeans.

FIGURE 50
Grain drier with a black plastic tube as solar collector. Air blown through the tube by a fan gains heat and is then forced through the grain bin. As external power is needed to run the fan, this drier may be uneconomical in many areas (158).

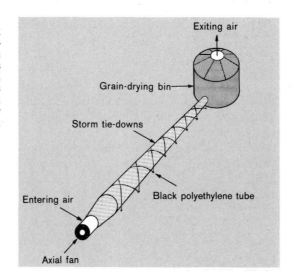

Solar cookers

Several small focusing collectors have been designed for using the sun's energy to cook food. A typical solar cooker consists of highly polished surfaces which reflect sunlight onto a small area. The reflectors concentrate solar radiation and intensify it to the energy level of an open fire (147). Two designs are shown in Figure 51.

Another cooking device is the solar oven, which is an insulated box with a transparent window on the side exposed to the sun. Collectors are arranged so as to reflect sunlight through the window and produce by intensification the energy equivalent of a fuel-fired oven (147).

Solar cookers generally have a collector area of about one square metre or less, a cooking area 20–30 centimetres in diameter and an effective cooking power of 0.25–0.50 kilowatt and can cook enough food for an average family. They have had little acceptance, however, because they often do not reach high enough temperatures for thorough cooking and because people living in hot climates rarely eat hot meals at midday.

Furthermore, as most developing countries cannot manufacture the polished aluminium, aluminized plastic and silvered mirror glass required to construct the reflectors, they would have to purchase them · from more industrialized nations (239). These materials also tend to be fragile and deteriorate after a few years. Local *in situ* construction of solar cookers is difficult owing to the precision needed in positioning

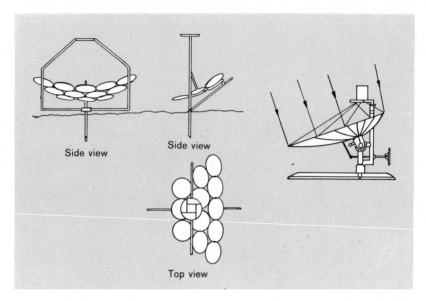

FIGURE 51. Solar cookers. On the left are three views of a cooker with twelve mirrors to collect sunlight and reflect it onto a small area. The mirrors are arranged parabolically so as to focus on a single spot (239). The cooker on the right has a single parabolic reflector with a focal length of 50 centimetres. It can cook a complete meal in two hours (205).

the reflectors. Moreover, the cost of simple solar cookers ($12–$35) is too high for most families in the developing countries (165). Another major drawback is that solar cookers can be used only under clear skies.

Even so, the potential of solar cookers may be realized if researchers can make use of readily available materials, simplify their construction and improve their durability. For example, in India a "sun basket" utilizing local available materials and village technologies has been developed to harness solar energy for household use. Essentially, the sun basket is a bamboo tokra. The inside is papier mâché lined with aluminium foil. The papier mâché (made of wheat flour, methi and shredded waste paper) and the bamboo are formed over a parabolic plaster-of-paris mould. After drying, the mould is removed, the inside of the basket smoothed and the aluminium foil pasted on it to form the parabolic mirror (272).

Solar cabinet

Spoilage in the food industry could be reduced substantially if simple low-cost driers were available (256). Experiments with the dehydration of grapes, chillies and dates in home-made solar cabinet driers developed in Jodhpur, India, have been successful (92). The use of solar cabinet driers could save on fuels and electricity consumed by the industry for dehydration (93, 94, 95). Furthermore, the increased earnings from cash crops and dehydrated fruits and vegetables would more than offset the minimum cost of applying this type of solar technology.

For direct drying, small produce can be exposed to solar radiation in a "hot box." The drying temperature of 50°–80°C inside the box is the result of the extraction and vaporization of the moisture from the product by solar radiation.

The solar cabinet drier is a rectangular box made of wooden planks (25 millimetres thick) with a base area of 1.5 square metres (93). The roof of clear window glass (3 millimetres thick) is inclined so as to receive maximum solar radiation year round. The floor of the drier is insulated with sawdust. Humid air escapes from numerous holes drilled in the sides and base of the cabinet, thus creating a partial vacuum which induces fresh air from the base. The entire inside of the drier is painted matte black for the maximum absorption of solar heat. The produce is placed in the drier through a rear door onto a removable wire-mesh screen (93).

Solar refrigeration

Many systems can be considered for solar refrigeration. As yet, the optimum scale of operation for solar coolers in developing countries is unknown. A better use of available foodstuffs would be possible if solar refrigeration could be successfully provided (73). It has been estimated that even in the temperate climate of France 25% of all food produced would be lost in the absence of refrigeration (91).

A solar refrigerator consists of two components: a solar-powered unit and a refrigeration unit. The power unit may be a flat plate collector or a focusing collector, and it can be operated as a continuous or intermittent absorption system. If electric power is needed to operate the pumps, the continuous absorption system is uneconomical for the developing nations. The intermittent absorption refrigeration system is preferred in rural areas where electricity is not available.

From the results of experiments the most promising cooling device for developing nations appears to be the ice maker (73). An economically viable solar ice maker for domestic or village use is likely in the near future (73). The larger village-size units would be more efficient and less costly.

Attempts are being made to design, construct and test a solar ice maker capable of producing 100 kilograms of ice daily without using oil or electricity. Such a unit would require a solar collecting surface of about 20 square metres (73). The commercial viability of this cooling unit will be assured provided it can manufacture about half a kilogram of ice for about one cent (U.S.). Solar refrigeration devices for domestic use must be as automatic as possible to compete with electric refrigerators.

Solar refrigeration using ammonia-water and ammonia-sodium thio-cyanate systems has been studied. During the refrigeration phase of water-ammonia systems the generator-absorber is allowed to cool. As the pressure drops, the evaporating ammonia draws heat from its surroundings. The evaporated ammonia is absorbed by the weak ammonia solution in the generator-absorber. The process proceeds until all the ammonia in the condenser has evaporated.

Although solar cooling units can be further improved, it will remain much easier to decrease humidity than to provide refrigeration. In tropical climates, where the demand for air cooling is high, air-conditioning operated by solar energy is technically feasible (107). Dehumidification of the air followed by evaporation of the water and restoration of a part of the humidity results in considerable cooling even in hot, humid climates (107).

Photovoltaic cells

Solar cells directly convert solar energy into electricity. Unlike other solar devices they do not rely on heat to perform useful work.

The development of photovoltaics is largely attributable to the U.S. space programme, which used solar cells on the Mercury and Apollo missions. Although the type used by NASA cost from $200 to $600 per watt (89), the actual structure of solar cells is simple and the chief component is made from sand.

Solar cells are based on the principle of the photoelectric effect. During the past century small quantities of energy which did not possess a charge were discovered. These were named photons. The basic photoelectric effect describes the release of an electron when a sufficiently energetic photon of light is absorbed by certain materials (158). Photovoltaic cells absorb photons from sunlight, create an electric field and act as a semiconductor to produce an electrical current.

The material most widely used in photovoltaic cells is silicon, which is abundant in sand as silicon dioxide (SiO_2). A solar cell consists of two waferlike layers of silicon separated by a thin barrier. The bottom layer of silicon is treated with a chemical, such as arsenic. The chemical bonding of the two materials results in excess electrons, thereby producing a negative-type silicon. The upper layer is treated with boron to produce a positive-type silicon, which lacks electrons. When struck by sunlight, the upper surface of the cell absorbs photons, which free

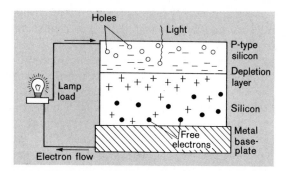

FIGURE 52. Solar cell. The two layers of silicon are separated by a thin barrier, or depletion layer. The upper layer of silicon is treated with chemicals to create a deficiency of electrons, and the bottom layer is processed to create an excess of electrons. The upper surface of the cell absorbs energy from solar radiation, which displaces electrons and forces them across the depletion layer. From there the free electrons flow through a wire, perform useful work and flow back to the top layer of silicon — ready to start another cycle (66).

electrons near the positive-negative junction, thus creating "holes" in their place. The freed electrons cross over the barrier, or depletion layer, and accumulate on the negative side, forming a voltage potential between the two silicon layers (Fig. 52). Wires connected to metal contacts allow the electrons to flow through a load from the negative layer back to the positive layer to perform work.

The high cost of solar cells results from the need for ultrapure silicon. Whereas a kilogram of 99% pure silicon only costs about $1.10, the 100% purity required for solar cells raises the processing cost to about $65 a kilogram (89). Nevertheless, in the past few years the price of photovoltaic cells has dropped from almost $600 to $20–$30 a watt (89). This reduction is significant, but the price is still too high to be competitive.

The U.S.A., Japan, the Federal Republic of Germany, France and the United Kingdom are conducting research on low-cost terrestrial photovoltaic power systems (39, 236, 237, 277). Most feel that costs can be reduced significantly, perhaps to only a few hundred dollars per average kilowatt, by 1985 (277). This might be accomplished by using materials other than silicon or by developing ways to use less pure substances. Also, the efficiency of only about 12% of current solar cells might be increased to as high as 30%, although an efficiency of between 11% and 23% appears to be more realistic (192).

The unique advantages of photovoltaic cells have prompted great interest in them, not only in industrialized nations but also in the developing nations. There are no waste products, they require little maintenance and they can operate indefinitely, it appears, without degradation. Solar cells seem to be ideal for small decentralized power sources. They could, for example, provide power for operating offshore lighting for navigation, telecommunication equipment, refrigerators in rural health dispensaries, small electrical appliances in remote hotels and motels, educational television receivers and audio-visual equipment in rural schools and community centres.

In the U.S.A., photovoltaic cells are used to operate lights and horns on offshore oil platforms, radio repeater stations and railway guide crossing equipment (278). In Japan, silicon solar cells have been used for more than a decade in telecommunication stations, lighthouses, buoys and robot rain gauges (9). In Pakistan, solar cell systems in kit form are available for lighting rural homes (9).

A photovoltaic system for the direct conversion of sunlight into electricity closely resembles a flat plate collector (Fig. 53). For small-scale applications the required solar cell is small. The solar power input at noon on a clear day at sea level is about one kilowatt per square metre. Therefore, a square metre covered by solar cells with a conversion efficiency of 10% would produce about 100 watts. Typical required power levels are about one watt for highway and radio fire-

FIGURE 53. Photovoltaic cell. The cell is encased in a boxlike cabinet. The side of the cabinet facing the sun is made of transparent material. Sunlight strikes the panel, as in a plate collector, and is converted into electricity (277).

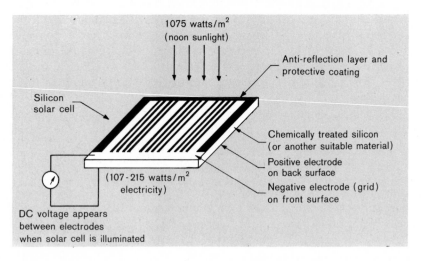

alarm call boxes, 24 watts for repeater stations for train control, and 50 watts for remote communication repeater stations (278). Solar cells can be added to a photovoltaic system to increase its power output.

Photovoltaic cells present many of the same problems of other solar devices. The need for electrical storage is just as acute as the need for heat storage with other solar applications.

Probably the most highly developed and efficient means of storing electricity is a lead-acid battery of the type used in automobiles. A typical battery consists of two lead electrodes, one of which is immersed in acid. When voltage is applied, electrons flow through a circuit from the lead-acid electrode to the lead electrode. The circuit can then be broken, and when it is reconnected, the electrons will flow back to the lead-acid electrode (55). In this way useful work is stored in the battery.

Commercial lead-acid batteries can store about 29 watt-hours per kilogram and occupy about 1.1 cubic metres per kilowatt-hour (55). Storage batteries for solar cells have an estimated life of six years, an overall efficiency of 85% and cost about 8.9 cents per volt-amp-hour of storage capacity (158).

The total cost of a photovoltaic system is $5.57 and $6.45 per kilowatt-hour with annual interest rates of 10% and 14% respectively (158). These calculations were made for a system generating an average of 3.3 watt-hours per day and having a 36-day storage capacity. This length of storage is assumed to be necessary to supply energy over the months when insolation is lower than the annual average.

Wind power [1]

The major factors responsible for the power output from a particular wind machine are area and wind speed. A minimum average wind speed of 10 kilometres per hour is required to operate a windmill. Doubling the area will double the power output, but doubling the wind speed will increase the power output to eight times the original. It is important that the wind machine be positioned so as to take best advantage of the prevailing wind.

The wind speed is affected by topography (hills and valleys) and by surface roughness (buildings and trees). Surface roughness tends to reduce wind speed near the ground (Fig. 54). As wind speed increases

[1] Adapted from AEIS 354 by Truman C. Surbrook and Jeffrey E. Friedle, Agricultural Engineering Department, Michigan State University, September 1976.

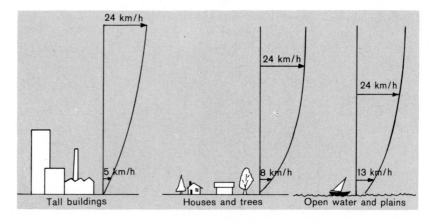

FIGURE 54. Effect of tall objects on wind speed near the ground.

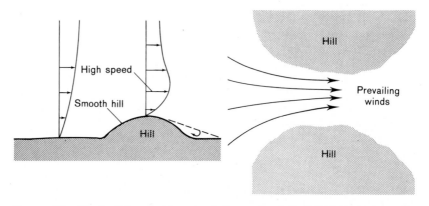

FIGURE 55. Effect of topography on wind speed. *Left:* Speed increases as the wind flows over a smooth hill. *Right:* Speed may be higher in a valley.

with elevation, a cost comparison should be made between increasing the rotor area and increasing the tower height to maximize the power output of the wind machine. Certain topographical features can augment wind speed. It will increase where wind is funnelled into a valley and will also be slightly higher near the ground at the top of a smooth hill (Fig. 55).

DETERMINING MAXIMUM POWER

The maximum power of the wind is determined by the area of the wind collector and the cube of the wind speed. This relationship in watts is expressed in the formula

$$Power_T = 0.01319 \times area \times (wind\ speed)^3$$

where

$Power_T$ = theoretical wind power in watts
area = useful area of rotor in square metres
wind speed = average in kilometres per hour

Unfortunately, it is not possible to extract all of the power in the wind. Actual power efficiency depends on the wind machine design (159). Four types of wind machines and their approximate efficiencies are listed in Table 44.

A further reduction occurs when shaft power is converted to electrical power owing to losses in gear boxes or pulley drives, in the alternator or generator and in other power conversion equipment. The conversion efficiency can vary from 50% to 80%. For the purpose of estimating power, 65% can be used in the formula

$$Efficiency_{system} = efficiency_{rotor} \times efficiency_{conversion}.$$

Realistic total system efficiencies for the conversion of theoretical wind energy into electrical energy lie between 10% and 30%.

The approximate electrical power in watts to be expected from a wind machine can be determined with the formula

$$Power = 0.01319 \times area \times (wind\ speed)^3 \times efficiency.$$

TABLE 44. – APPROXIMATE EFFICIENCY OF
VARIOUS WIND MACHINES

Type	Percent efficiency
High-speed propeller	45
Darrieus/Straw rotor	32
Multiblade turbine	30
Savonius rotor	16
Modified...........	18

FIGURE 56. Savonius rotor. The useful wind-collecting area is the radius times the height.

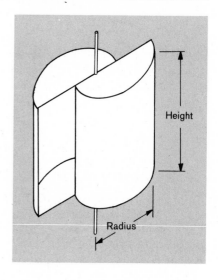

The area in the formula is the useful area of the rotor perpendicular to the wind direction. For the high-speed propeller and the multi-blade turbine types the area is easily determined by using the formula

$$\text{Area}_{\text{propeller}} = \frac{\pi}{4} \times (\text{diameter})^2 = 0.785 \times (\text{diameter})^2.$$

However, with the Savonius type only about half the area swept by the turning rotor is useful (Fig. 56). Thus

$$\text{Area}_{\text{Savonius}} = \text{radius} \cdot \times \text{height}.$$

The actual area perpendicular to the wind direction must be determined for the Darrieus/Straw rotor.[1]

For example, the potential power of a high-speed propeller wind machine 2 metres in diameter located in an area where the wind speed averages 24 kilometres per hour is calculated as follows:

Area = $0.785 \times (\text{diameter})^2 = 0.785 \times (2)^2 = 3.14$ m²
Wind speed = 24 kilometres per hour
Efficiency = efficiency$_{\text{rotor}} \times$ efficiency$_{\text{conversion}} = 0.45 \times 0.65 = 0.29$
Power = $0.01319 \times 3.14 \times (24)^3 \times 0.29 = 166$ watts

---

[1] The Darrieus rotor is here identified as the Darrieus/Straw rotor because a patent in the name of Straw predates that of Darrieus.

ROTOR WIND MACHINES

The rotors of wind machines are classified as horizontal or vertical depending on the orientation of the axis of rotation to the direction of the wind. The axis of a horizontal rotor is parallel to the wind stream, and some means of keeping the rotor faced into the wind is needed. The gyroscopic effect of the rotor results in slow tracking, and quick changes in wind direction (gusts) cannot be utilized. The axis of a vertical rotor, being perpendicular to the earth and the wind stream, accepts wind from any direction and can thus better utilize the extra power from gusts.

Rotors can furthermore be the drag or lift type. Drag rotors have lower efficiencies but develop higher torque, or turning power, at low speeds. Lift rotors produce power from both drag and lift forces owing to the airfoil shape of their blades. Lift rotors therefore generally operate at higher efficiencies and develop more power per unit area.

Savonius rotor

The Savonius rotor is a vertical-axis, drag- and lift-type wind machine. The slow-turning rotor has a high-starting torque and is well suited for pumping water.

Darrieus/Straw rotor

The Darrieus/Straw rotor is a vertical-axis, lift-type wind machine. The rotor has two or three blades with an airfoil cross section. Its maximum efficiency is 38%, but because of its low-starting torque it is not self-starting. A Savonius rotor could be added, as shown in Figure 57, to make it self-starting. It is a reasonably high-speed rotor and suitable for powering a generator.

Multiblade turbine

The multiblade turbine (Stewart Mill) is illustrated in Figure 58. The rotor has 15–40 blades on a horizontal axis. As this lift-type rotor operates at a relatively low speed, rotor balance is not critical. Because of its high-starting torque, the multiblade turbine is well suited for pumping water (116, 140).

High-speed propeller

The high-speed propeller is of the lift type and has two or three airfoil blades (Fig. 59). A three-blade propeller is easier to balance. The maximum efficiency for this type is 47%; despite the low-starting torque it is well suited for generating electricity.

FIGURE 57
Darrieus rotor. This three-blade rotor is often used to power wind generators. Small Savonius rotors are added at the top and bottom to make the Darrieus rotor self-starting.

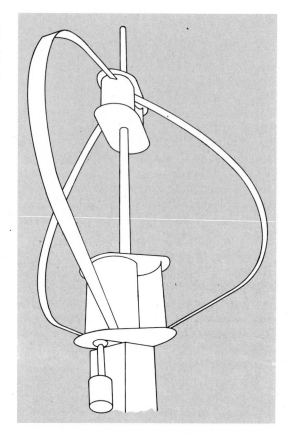

The rotor of both the high-speed propeller and the multiblade turbine must face into the wind. A tail vane, or rudder, is often used. Moreover, some means of braking the rotor or of turning it out of the wind must be provided so as to prevent damage to the rotor or other components when the wind speed becomes excessive.

WATER PUMPING SYSTEMS

A wind machine for pumping water consists of a rotor on a tower and a pump. The rotary motion is changed to a reciprocating motion to drive the pump piston. A rotor with high-starting torque is needed to overcome the large load of a piston pump (18).

Figure 60 shows maximum pumping rates with average wind (24-29 km/h) and rotors of different diameters. The pumping rate will vary as shown in Figure 61 owing to changes in velocity. A maximum

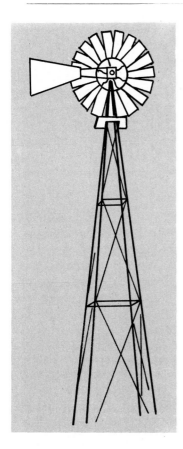

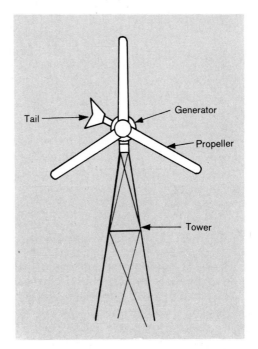

Left: FIGURE 58. Multiblade turbine. It is ideal for pumping water.

FIGURE 59. Three-blade high-speed propeller. It is widely used for powering wind generators.

FIGURE 60. Typical maximum pumping rates for windmills of different diameters at an overall efficiency of 10%. The wind speed was 24 kilometres/hour for small mills and 29 kilometres/hour for large mills.

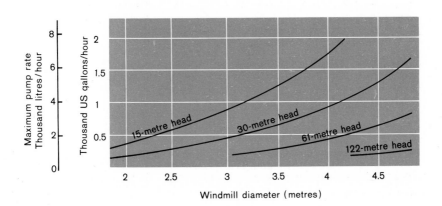

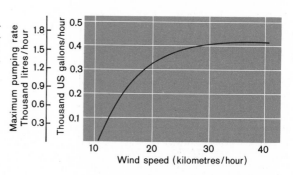

Figure 61
Typical maximum pumping rates for a 3-metre-diameter windmill and 30-metre head at various wind speeds.

rate is achieved because the rotor is designed to turn out of excessive winds. Water storage must also be provided for periods of little or no wind. In rural areas of the developing countries, pumping water for irrigation constitutes a major use of energy. Because windmills for pumping water are relatively inexpensive to build, they deserve more attention (128, 198).

ELECTRICAL GENERATING SYSTEMS

A major drawback of wind-generated electricity is the inconstant power supply (128). Before wind-generated electricity can be used for domestic purposes, certain adjustments in energy-use patterns are necessary. Activities requiring electrical energy must be scheduled so as to take advantage of the wind.

Most electrical appliances require alternating current with the constant voltage and frequency of utility power to prevent equipment damage. The voltage must be 115 or 230 and the frequency 50 or 60 hertz. The output voltage and frequency of the wind generator change with wind speed, but there are several solutions to this problem.

One method of voltage and frequency regulation is to use a direct-current generator to charge the batteries. An inverter is then used to produce alternating current of 115 or 230 volts and 50 or 60 hertz from the batteries even when the wind is not blowing. Only deep-discharge or deep-cycle batteries can withstand the extremes of charging and discharging over long periods. Regular car or truck batteries are subject to plate decomposition or buckling under deep discharge. Batteries must be stored in a warm area for maximum efficiency. A half-charged battery will freeze at —21°C, and a completely discharged battery will freeze at —8°C. The disadvantages of this method are the high cost and maintenance of the batteries.

A better approach is to use an inverter capable of producing 115-

or 230-volt, 50- or 60-hertz alternating current directly from the output of the direct-current generator powered by the wind machine.

A third approach is to use the output of the wind generator directly to perform a task not requiring constant voltage or frequency. One such task is heating water.

The total monthly output in kilowatt-hours of a particular generator can be estimated by multiplying the generator size in kilowatts by the appropriate factor in Table 45 for average wind speed. For example, a 1-kilowatt generator operating in an area where the average wind speed is 22 kilometres per hour will produce about 100 kilowatt-hours of electrical energy.

TABLE 45. – FACTORS FOR ESTIMATING MONTHLY ELECTRIC POWER OUTPUTS OF WIND-POWERED GENERATORS

Average wind speed (km/h)	Factor
13	40
16	60
19	80
22	100

Energy needs in fertilizer production

Fertilizers fall into two categories: organic and inorganic. Organic fertilizers are primarily crop residues and manure. Inorganic fertilizers are man-made compounds based on three essential elements: nitrogen, potassium and phosphorus. The worldwide use of 80–90 million metric tons of inorganic fertilizers represents a substantial monetary investment. World demand is expected to increase by 6% to 7% per year until 1980/81; in the developing countries an 11% annual increase is projected (123).

Table 46 compares the projected consumption and production of nitrogenous and phosphatic fertilizers by region for 1980/81 with the corresponding data for 1971/72. The developing countries produced a small fraction of the world's fertilizer supply in 1971/72 and consequently had to import about 5 million metric tons of nitrogen and

TABLE 46. – PROJECTED CONSUMPTION AND PRODUCTION OF NITROGENOUS AND PHOSPHATIC FERTILIZERS BY REGION FOR 1980/81 COMPARED WITH CORRESPONDING DATA FOR 1971/72

Region	1971/72			1980/81		
	N	P	Total	N	P	Total
	.... 10^6 *metric tons of plant nutrients*					
Developed regions						
Consumption	24.1	17.3	41.4	40.3	24.7	65.0
Production	29.7	19.6	49.3	35.1	18.9	54.0
Surplus or deficit	+5.6	+2.3	+7.9	—5.2	—5.8	—11.0
Developing regions						
Developing Asia						
Consumption/demand	4.0	1.4	5.4	11.0	4.4	15.4
Production	2.3	0.7	3.0	4.6	1.5	6.1
Surplus or deficit	—1.7	—0.7	—2.4	—6.4	—2.9	—9.3
Latin America						
Consumption/demand	1.4	1.0	2.4	3.0	2.0	5.0
Production	0.8	0.5	1.3	1.9	0.8	2.7
Surplus or deficit	—0.6	—0.5	—1.1	—1.1	—1.2	—2.3
Africa						
Consumption/demand	0.4	0.3	0.7	1.6	0.7	2.1
Production	0.1	0.4	0.5	0.6	0.9	1.5
Surplus or deficit	—0.3	+0.1	—0.2	—0.8	+0.2	—0.6
Subtotal for developing countries						
Consumption/demand	5.8	2.7	8.5	15.5	7.1	22.6
Production	3.2	1.6	4.8	7.1	3.3	10.4
Surplus or deficit	—2.6	—1.1	—3.7	—8.4	—3.8	—12.2
Asian centrally planned economies						
Consumption/demand	3.4	1.1	4.5	7.2	1.9	9.1
Production	2.1	1.1	3.2	1.9	1.1	3.0
Surplus or deficit	—1.3	–	—1.3	—5.3	—0.8	—6.1
Total for developing regions						
Consumption/demand	9.2	3.8	13.0	22.6	9.0	31.6
Production	5.3	2.7	8.0	9.0	4.4	13.4
Surplus or deficit	—3.9	—1.1	—5.0	—13.6	—4.6	—18.2
World total						
Consumption	33.3	21.1	54.4	62.9	33.7	96.6
Production	35.0	22.3	57.3	44.1	23.3	67.4
Surplus or deficit	+1.7	+1.2	+2.9	—18.8	—10.4	—29.2

SOURCE: (123)

phosphorous fertilizer; by 1980 they will have to import over 18 million metric tons.

ORGANIC FERTILIZERS

For centuries man has been using organic matter to maintain or increase soil productivity. Manure and fertilizer were practically synonymous terms (143). In China the land has been cultivated without a loss of productivity for thousands of years under a careful management system (151). There, and elsewhere, the use of nitrogen-fixing legumes (see pages 141–146) and the application of animal and human wastes are traditional methods of maintaining the nutrient balance in the soil.

Animal and human excrement contains elements essential to plant growth. The specific nutrient content of animal excrement depends on the species and the nutrient content of feed. In the U.S.A., manure generally contains 31 kilograms of nitrogen, 6.2 kilograms of phosphorus and 22 kilograms of potassium per dry metric ton (8). Table 47 compares the nutrient content of manure and inorganic fertilizer, and

TABLE 47. – COMPARISON OF NUTRIENT CONTENTS OF MANURE AND FERTILIZER, U.S.A., 1973

Nutrient	Total (10^3 metric tons)	Nutrient content	
		Kilograms per dry metric ton	Pounds per wet English ton
Nitrogen			
Manure	7 367	31	10
Fertilizer	7 540	191	—
Phosphorus			
Manure	1 463	6.2	2
Fertilizer	2 222	56	—
Potassium			
Manure	5 342	22	7
Fertilizer	3 836	97	—

SOURCE: (143)

specific data on excrement production are given in Table 48. Human excrement contains the highest percentage of nitrogen, and horse, mule and donkey manure contains the least.

Table 49 shows the findings from case studies carried out in several villages to determine the nitrogen fertilizer potential of their plant, human and animal wastes. Other than in Arango, Mexico, and Peipan, China, more nitrogen than was available from local wastes was currently being used. Proper management and use of organic wastes could provide a significant amount of nitrogen, but not enough for high-yielding seed varieties (151). Tables 50 and 51 depict the yield response of rice and wheat to increasing nitrogen inputs. (The areas listed in Table 49 could likewise increase their yields by a few hundred kilograms per hectare with efficient use of wastes.)

TABLE 48. – EXCREMENT PRODUCTION DATA

	Production per 1 000 kg liveweight (kg/yr)	Assumed average liveweight (kg)	Production per head (kg/yr)	Moisture content (percent)	Nitrogen content (percent of dry matter)	
					Solid and liquid wastes	Solid wastes
Cattle	27 000	200	5 400	80	2.4	1.2
Horses, mules, donkeys	18 000	150	2 700	80	1.7	1.1
Pigs	30 000	50	1 500	80	3.75	1.8
Sheep and goats .	13 000	40	500	70	4.1	2.0
Poultry	9 000	1.5	13	60	6.3	6.3
Human faeces without urine .	—	40–80	50–100	66–80	—	5–7
Human urine ...	—	40–80	18–25 (dry solids)	—	15–19 (urine only)	—

SOURCE: (104)

TABLE 49. – ANNUAL ORGANIC NITROGEN SUPPLY IN CASE-STUDY VILLAGES

	Hectares of cultivated land	Excluding human excrement			Nitrogen collectable in solid and liquid human excrement [2]	Total nitrogen collectable in solid and liquid wastes	Nitrogen available per hectare cultivated land
		Nitrogen available in net residues [1]	Nitrogen in collectable solid wastes	Nitrogen in collectable solid and liquid wastes			
		.... *Kilograms*					
Mangaon, India	300	10 000	4 000	7 000	3 000	10 000	33
Peipan, China [3]	200	13 000	7 000	9 000	4 000	13 000	65
Kilombero, Tanzania	60	1 000	500	800	300	1 100	18
Batagawara, Nigeria	530	13 000	6 000	9 000	4 200	13 200	25
Arango, Mexico [3]	380	7 000	5 000	6 000	1 700	7 700	20
Quebrada, Bolivia (one *parcela*) .	1	70	40	50	18	68	68

SOURCE: (151)

[1] Assuming a nitrogen content of 0.2% for crop residues. – [2] Assuming that a maximum of 80% of the nitrogen in human excrement can be collected. For Arango and Peipan, intermediate values for the total nitrogen excreted per person per year have been used, and for the other villages, lower figures have been used. The rationale is that people in Arango and Peipan have more adequate diets, which produce excrement with a higher nitrogen (protein) content. – [3] The nitrogen figures for Peipan and Arango are probably underestimates as crops there are well fertilized. The nitrogen content of crop residues is probably considerably higher, especially because maize, which has a higher nitrogen content in the residue, is a major crop in both areas.

TABLE 50. – RICE YIELD RESPONSES TO NITROGEN INPUTS IN THE PHILIPPINES, 1970[1]
(kilograms per hectare)

Quantity of nitrogen	Rice yield	Marginal product per 15 units
0	4 900	—
15	5 250	350
30	5 600	350
45	5 900	300
60	6 150	250
75	6 400	250
90	6 650	250
105	6 850	200
120	7 050	200
135	7 200	150
150	7 250	50
165	7 300	50
180	7 300	0

SOURCE: (50)

[1] The relationship between nitrogen inputs and yields is not linear.

TABLE 51. – WHEAT YIELD RESPONSES TO NITROGEN INPUTS IN INDIA
(kilograms per hectare)

Quantity of nitrogen	Wheat yield	Marginal product per 20 units
0	3 844	—
20	4 356	512
40	4 818	462
60	5 230	412
80	5 592	362
100	5 904	312
120	6 166	262
140	6 377	211
160	6 539	162
180	6 651	112
200	6 712	61
220	6 724	12
240	6 685	—39

SOURCE: (50)

INORGANIC FERTILIZERS

Nitrogen

Nitrogen (N) is a colourless, odourless gas that comprises about 80% of the earth's atmosphere. Atmospheric nitrogen is not readily available to most vegetation, but it can be combined with other elements to produce a compound that can be applied to soil and used by plants.

Nitrogenous fertilizers are produced by combining hydrogen with atmospheric nitrogen to form synthetic ammonia (264) — the base for at least 90% of all nitrogen fertilizers (19). This process is represented by the following chemical equation:

$$2N + 3H_2 \longrightarrow 2NH_3$$

In the U.S.A. the hydrogen is obtained from natural gas — methane (CH_4). In Europe fuel oil, coal and naphtha are widely used as a hydrogen feedstock. Norway has been using hydrolized water (264). Typically, steam is used to reform the hydrocarbons from feedstocks into hydrogen and carbon dioxide (CO_2). After purification the hydrogen is combined with atmospheric nitrogen to produce ammonia (264).

In the U.S.A. about 38% of the anhydrous ammonia produced is used directly as fertilizer, and the remainder is used to produce urea, ammonium nitrate or compound fertilizers (264). Urea is fast becoming the world's major source of solid nitrogen. Whereas in 1967 urea production was about 19% of total nitrogen production, in 1973 it was about 30% of the total (19). The increase is due to its high analysis (46% N) and the absence of fire and explosion hazards (19). To produce urea, anhydrous ammonia is combined with carbon dioxide as follows:

$$2NH_3 \text{ (liquid)} + CO_2 \text{ (gas)} \xrightarrow[\text{heat}]{\text{pressure}} CO(NH_2)_2 \text{ (urea-solid)} + H_2O$$

Table 52 shows the worldwide consumption of oil and natural gas for nitrogen fertilizers. About 40 million metric tons of nitrogen fertilizer costing about $8 thousand million were used in 1974 (108). In 1975 the developed regions used almost four times as much oil and natural gas for nitrogen fertilizer production as the developing regions did; and, accordingly, the developed regions applied four times as much nitrogen fertilizer per hectare of cereal cropland (Table 53).

The greater use of fertilizers in the developed nations has increased yields per hectare. In 1971 the developing countries applied about a third as much nitrogen fertilizer per area under cereal cultivation as the developed countries did. As a result, yields in the developing countries were about 1 000 kilograms per hectare less (Fig. 62). (Note, however, that while fertilizer use differed by a factor of three, yields differed by a factor of less than two.)

Experiments with maize show optimum yields with the application of 225 kilograms of nitrogen per hectare. Yet, the optimum kilocalorie output per input is seen to occur when about 135 kilograms of nitrogen are applied per hectare (Fig. 63). Farmers must therefore choose between optimum yields and optimum input/output energy ratios.

Phosphate

Phosphorus compounds comprise the second largest group of fertilizers. The consumption and production of phosphate fertilizers

TABLE 52. – ENERGY CONSUMPTION OF OIL AND NATURAL GAS FOR THE MANUFACTURE OF NITROGEN FERTILIZERS BY WORLD REGIONS, 1975

Region	Consumption (10^{16} joules)
North America (U.S.A.)	52
Western Europe	40
Eastern Europe and the U.S.S.R.	56
Oceania	1
Other developed countries (Israel, Japan, South Africa)	6
Total for developed regions	155
Latin America	10
Africa (excluding South Africa)	5
Asia (excluding Japan and Israel)	24
Total for developing regions	39
Other Asia	24
TOTAL	218

SOURCES: (29, 247)

increased about 150% between 1960 and 1973 — considerably less than the increase for nitrogen (264). Worldwide use, approximately 25.8 million metric tons in 1973, is expected to range from 31.2 million metric tons to 37.6 million metric tons by 1980 (264). (See Fig. 64 and Table 46.)

The basic method of producing phosphate fertilizers is the decomposition of phosphate rock with sulphuric, phosphoric or nitric acid. Sulphuric acid is efficient (over 90% of the phosphorus is recovered), but large quantities of sulphate wastes are produced. The use of nitric acid requires further treatment, usually with ammonia, to produce a suitable fertilizer product. Phosphoric acid is combined with potash and/or ammonia to produce ammonium phosphates or potassium phosphates (264).

TABLE 53. – NITROGEN FERTILIZER APPLICATIONS AND COSTS FOR CEREAL CROPS BY WORLD REGIONS, 1975

Region	Metric tons per hectare	Cost per hectare at $700/ metric ton
North America (U.S.A.)	0.12	84
Western Europe	0.16	112
Eastern Europe and the U.S.S.R.	0.075	52.5
Oceania	0.019	13.3
Other developed countries ...	0.11	77
Average for developed regions	0.10	70
Latin America	0.042	29
Africa (excluding South Africa)	0.015	10.5
Asia (excluding Japan and Israel)	0.024	16.8
Other Asia	0.037	25.9
Average for developing regions	0.025	17.5
WORLD AVERAGE	0.059	41.3

SOURCES: (29, 247)

The manufacture of phosphate fertilizers requires 126 to 200 $\times$ 10^8 joules per metric ton of P_2O_5, compared with 517 to 654 $\times$ 10^8 joules per metric ton for nitrogen fertilizers (84).

There are at least 90 $\times$ 10^8 metric tons of phosphate rock in the world. As the 1972 production was only about 13 $\times$ 10^6 metric tons (264), the extensive world phosphate reserves should not be depleted for a long time (Table 54). However, environmental problems in the mining of phosphate rock are mounting, as most reserves are strip-mined, causing erosion and temporary destruction of land for other uses.

World sulphur resources must also be considered, as phosphate fertilizer production requires large quantities of sulphuric acid (Table 55). Sulphur is obtained either from ore deposits or recovered as a by-product of other operations. The current practice of obtaining

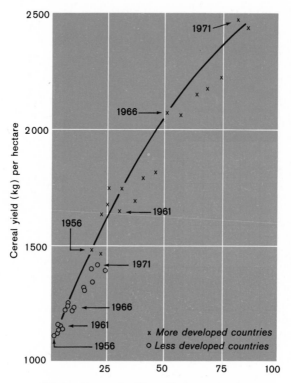

Total nitrogen fertilizer (kg) per hectare
under cereal cultivation

FIGURE 62. Relationship between the use of nitrogen fertilizers and the yield of cereal grains in developed and developing countries, 1956–71. Note that the total nitrogen fertilizer use on all cultivated land is divided by the area under cereal grains; thus the actual application rates would be about one half the rates shown here as only one half of the total land area is under cereal grains (110).

sulphur primarily from cheap ores is expected to change drastically because the world reserves are being depleted and the sulphur discarded as a waste product from other operations is posing environmental problems (264). The long-range expectation is that 93% of all sulphur will be obtained by costly removal from solid, liquid and gaseous effluents of co-product operations. Currently, 72% is obtained from sulphur ore deposits (264).

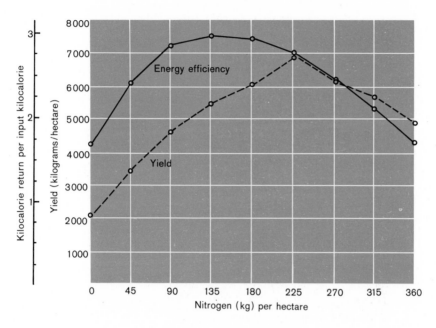

FIGURE. 63. Yield and kilocalorie return per input kilocalorie for maize at different rates of nitrogen fertilizer application. Maize showed optimum yields with the application of about 200 kilograms of nitrogen fertilizer per hectare, whereas the optimum kilocalorie return per input kilocalorie resulted from an application of about 135 kilograms of nitrogen fertilizer per hectare (190).

FIGURE 64. World fertilizer consumption. In 1973, total consumption was over 77 million metric tons, which represented more than a fivefold increase from 1950. The use of inorganic fertilizers is expected to increase significantly in the years ahead (241).

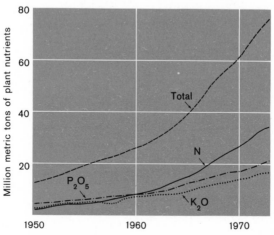

TABLE 54. – ASSESSMENT OF WORLD PHOSPHORUS RESOURCES (PHOSPHATE ROCK AND APATITE) RECOVERABLE AT VARIOUS PRICES BY REGION AND COUNTRY

	Price per metric ton of phosphorus in 1972 dollars		
	$64 [1]	$95	$159
	.... × 10^6 *metric tons P*		
North America			
U.S.A.			
Florida	152	202	317
North Carolina	34	89	253
Tennessee	4	15	76
Idaho	22	253	760
Montana	1	33	152
Utah	24	51	291
Wyoming	1	18	55
Other states	—	13	58
Other countries	1	4	127
Total	239	678	2 089
South America	20	65	722
Europe			
U.S.S.R.	114	285	570
Others	5	8	16
Total	119	293	586
Africa			
Algeria	18	42	83
Morocco	68	683	3 415
Spanish Sahara	56	98	141
Tunisia	35	70	105
Others	60	252	500
Total	237	1 145	4 244
Asia			
China	8	34	134
Viet Nam	10	21	55
Israel	5	20	53
Jordan	14	26	...
Others	8	54	353
Total	45	155	595
Oceania			
Australia	—	76	165
Christmas Island	12	16	—
Others	6	8	—
Total	18	100	165
WORLD TOTAL [2]	678	2 436	8 401

SOURCES: (261, 264)

[1] 1972, 70 BPL, value $57.64 f.o.b. U.S. plants. – [2] World production in 1972 was 13 · 10^6 metric tons.

Price per metric ton of sulphur in 1972 dollars

	Mined[1]				Recovered[2]				Total			
	$17.50	$27.50	$37.50	$47.50	$17.50	$27.50	$37.50	$47.50	$17.50	$27.50	$37.50	$47.50
					10^6 metric tons							
North America												
U.S.A.	41	142	722	1 799	36	96	158	158	77	238	880	1 957
Canada	5	10	371	1 092	386	793	1 194	1 194	391	803	1 565	2 286
Mexico	10	41	132	310	5	15	20	25	15	56	152	335
Others	–	–	20	66	–	–	–	–	–	–	20	66
Total	56	193	1 245	3 267	427	904	1 372	1 377	483	1 097	2 617	4 644
South America	10	21	137	376	20	56	91	91	30	77	228	467
Europe												
U.S.S.R.	15	36	152	371	15	36	51	61	30	72	203	432
Poland	15	36	76	127	–	5	15	20	15	41	91	147
France	–	–	96	295	21	63	66	66	21	63	162	361
Spain	10	20	107	279	–	–	–	–	10	20	107	279
Italy	5	10	76	208	–	5	5	5	10	15	81	213
Germany (F.R.)	–	–	36	97	5	10	15	20	5	10	51	117
Finland	5	10	15	20	–	–	–	–	5	10	15	20
Others	10	20	163	462	5	15	31	36	15	35	194	498
Total	60	132	721	1 859	46	134	183	208	106	266	904	2 067
Africa	5	10	142	406	10	20	31	31	15	30	173	437
Asia												
China	5	10	76	203	31	46	56	61	36	56	132	264
Japan	10	20	56	132	31	51	66	81	41	71	122	213
Near East	5	10	87	234	467	691	777	782	472	701	864	1 016
Others	–	–	71	208	30	51	61	66	30	51	132	274
Total	20	40	290	777	559	839	960	990	579	879	1 250	1 767
Oceania	–	–	51	158	5	15	20	26	5	15	71	184
WORLD TOTAL	151	396	2 586	6 843	1 067	1 968	2 657	2 723	1 218	2 364	5 243	9 566

SOURCES: (264, 277)

[1] Mined sulphur is obtained from ores. – [2] Recovered sulphur is obtained from by-product wastes of other manufacturing processes. The use of recovered sulphur is expected to increase greatly in the near future.

Potassium (potash)

About 95% of the world output of potash comes from underground mines (264). There are two methods of obtaining potassium (K) from bedded ore deposits. The ore may be broken up and hauled to the surface, where it is mixed with a salt solution of sodium and potassium chlorides. After heating, purification and cooling, the potassium dissolves and eventually crystallizes as 99% pure potassium chloride (264).

Another method, used in some Canadian mines, involves the drilling of holes down to the potash bed. Water pumped through the holes into the bed dissolves the ore and forms a sodium-potassium brine which is subsequently pumped back to the surface. Potash manufacture requires about 63×10^8 joules per metric ton (84).

Large potassium deposists are isolated in a few areas. Canada possesses almost half of the world's potassium resources followed by Germany (D.R. and F.R.) and perhaps by the U.S.S.R., although the extent of deposits there is uncertain (Table 56).

Worldwide production of potash exceeded consumption in 1974 by an encouraging amount (Table 57); in fact, from 1960 to 1973 production increased 147% as compared with a 125% increase in consumption (164). Table 58 shows, however, that demand is likely to surpass production by 1980 if the current capacity is not increased (164). Canadian producers, the dominant exporters of potash, have indicated a willingness to expand production to meet world demand.

ADVANTAGES, DISADVANTAGES AND PROBLEMS
OF ORGANIC AND INORGANIC FERTILIZERS

Considering the rapid increase in the use of inorganic fertilizers and the large quantity of energy needed to produce them, it is beneficial to point out the differences between organic and inorganic fertilizers and the advantages and disadvantages of both (Table 59).

The nutrient content of fertilizers is shown in Table 47. Manure has a considerably lower content of nitrogen, potassium and phosphorus than manufactured fertilizers do. However, manure, straw and other organic materials contain other minerals generally not found in inorganic fertilizers and tend to improve the soil's ability to hold water, its crumb structure and its resistance to erosion by water and to crusting in beating rain (6, 125).

The choice between organic and inorganic fertilizers calls for consideration of some important indirect factors as well. Human and animal wastes require safe disposal as they often contain disease-carrying

TABLE 56. – ASSESSMENT OF WORLD POTASSIUM RESOURCES (PRINCIPALLY SYLVITE) RECOVERABLE AT VARIOUS PRICES BY REGION AND COUNTRY

	Price per metric ton in 1972 dollars		
	$53	$61	$66
	.... 10^6 *metric tons K*		
North America			
U.S.A.	90	268	268
Canada	9 070	19 300	19 300
Others	1	5	5
Total	9 161	19 573	19 573
South America			
Chile	9	18	27
Others	18	36	55
Total	27	54	82
Europe			
France	45	182	181
Germany (D.R.)	2 310	6 030	6 020
Germany (F.R.)	2 270	5 730	5 730
Italy	27	45	72
Spain	45	182	181
U.S.S.R.	4 540	9 070	12 000
United Kingdom	45	91	181
Others	63	172	236
Total	9 345	21 502	24 601
Asia			
Israel and Jordan	213	1 150	1 170
Others	5	9	9
Total	218	1 159	1 179
Africa			
Congo	18	45	91
Oceania	9	9	9
WORLD TOTAL [1]	18 778	42 342	45 535

SOURCES: (261, 264)

[1] World production in 1972 was 16 × 10^6 metric tons.

TABLE 57. – ESTIMATED WORLD POTASH PRODUCTION, DEMAND AND BALANCE, BY REGION, 1974

	Production	Consumption	Balance
	.... 10^3 *metric tons* K_2O		
North America	825	467	358
Western Europe	527	511	16
Eastern Europe and the U.S.S.R.	700	651	49
Japan	0	68	—68
Others	87	36	51
Total for developed regions	2 139	1 733	406
Latin America	0	96	—96
Developing Africa	34	17	17
Developing Asia	0	72	—72
Total for developing regions	34	185	—151
Other Asian	10	23	—13
WORLD TOTAL	2 183	1 941	242

SOURCE: (261)

organisms. Using these wastes as fertilizer can provide a convenient sanitary solution to the disposal problem.

Fertilizers are becoming significant environmental problems as regards air and water pollution. The manufacture of phosphate fertilizers results in substantial emissions of particulates, fluorides and sulphur oxides (84). Water pollution can be caused by both organic and inorganic fertilizers. Rain water and runoff leach the nutrients out of the fertilizer and carry them to the water table. Nitrates are the greatest pollution threat because they promote algae growth in lakes and streams and contaminate water supplies (6). Excessive fertilizer use can cause toxic nitrate levels. Another drawback of inorganic fertilizers is the contribution of the manufacturing process to water pollution.

It is questionable whether traditional organic fertilizers can supply the nutrients required in the intensive agricultural production of the developed countries. In Europe 171.5 kilograms of inorganic fertilizer

TABLE 58. – ESTIMATED AND PROJECTED WORLD POTASH FERTILIZER PRODUCTION AND CONSUMPTION BY REGION, 1978 AND 1980

	1978 produc- tion capacity	1978 produc- tion [1]	1980 Consumption		
			High	Mid-point	Low
.... *10⁴ metric tons K_2O*					
North America	942	825	630	591	552
Western Europe	671	587	631	594	556
Eastern Europe and the U.S.S.R.	1 020	889	969	896	822
Japan	0	0	88	74	61
Other developed nations [2] .	74	66	54	45	38
Total for developed regions	2 707	2 367	2 372	2 200	2 029
Latin America	0	0	206	185	165
Developing Africa	45	34	30	26	22
Developing Asia	0	0	126	117	108
Total for developing regions	45	34	362	328	295
Other Asian countries [3] ...	12	10	32	32	24
WORLD TOTAL	2 764	2 411	2 766	2 560	2 348

SOURCE: (264)

[1] Based on plants in developed countries operating at 95% of capacity and in the developing countries at 70% of capacity. – [2] Includes South Africa, Israel and Oceania. – [3] Includes China, the Democratic People's Republic of Korea, Viet Nam and Mongolia.

were applied per hectare in 1970 (Fig. 65), compared with only 7.6 kilograms per hectare in Africa and between 24 and 70 kilograms in the rest of the world. Nevertheless, the total quantity of fertilizer applied is significant and by 1980 is expected to increase 95% (87).

BIOLOGICAL NITROGEN FIXATION

Nitrogen cycle

Nitrogen is essential for plant growth. A colourless, odourless gas, it comprises about four fifths of the earth's atmosphere. Because most plants cannot use nitrogen in gaseous form, nature has devised

TABLE 59. – ADVANTAGES AND DISADVANTAGES OF ORGANIC AND
INORGANIC FERTILIZERS

Organic	Inorganic
Large non-nutrient content	High concentration of nutrients
Bulky	Ease of transport and handling
Little direct cost	Increasing cost
Largely renewable	Made from finite resources
Imprecise content analysis	Precise content analysis
No direct energy use in manufacture	Large direct energy use in manufacture
Readily available	Availability depends on production, cost and region
Provides disposal of wastes	Creates wastes in processing, but can also utilize wastes from other manufacturing processes

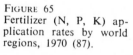
FIGURE 65
Fertilizer (N, P, K) application rates by world regions, 1970 (87).

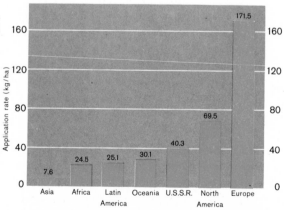

a complex system for converting it into a usable form (149). The natural processes for changing nitrogen into usable form produce about 220 × 10⁶ metric tons per year, far more than the 40 × 10⁶ metric tons of nitrogen produced by world industries in 1974 (108).

Abiological converters include lightning, combustion and ozonization. These fix about 10, 20 and 15 × 10⁶ metric tons of nitrogen, respectively, each year (27). Biological processes convert perhaps 175 × 10⁶ metric tons per year, of which about 90 × 10⁶ metric tons are fixed in cultivated soil (Table 60). Because the data are limited, these figures should be considered rough estimates only.

The nitrogen cycle is dynamic. Atmospheric nitrogen is converted into a compound, such as ammonia, which plants can absorb through their roots. The nitrogen compound is then incorporated into amino acids and proteins. The amino acids and proteins are eventually returned to the soil through the decomposition of plants or by the animals which have eaten the vegetation. Further decomposition converts the amino acids into nitrates that are used by new plants or converted by denitrifying bacteria into nitrogen, which escapes into the atmosphere.

Biological nitrogen fixation is the reduction of atmospheric nitrogen to ammonia by bacteria and blue-green algae. These microorganisms either live free around the roots of higher plants, forming a loose association with them, or share a symbiotic (interdependent) relationship with a higher plant. Symbionts are associated with only a few

TABLE 60. – SOURCES OF NITROGEN FOR CROP PRODUCTION, 1975

Source	10⁶ metric tons
Industrial fixation (fertilizer)	42
Biological fixation — total	175
Agricultural soil	90
Legume crops	40
Non-legume crops	9
Meadows and grasslands	45
Lightning	10
Combustion	20
Ozonization	15

SOURCE: (281)

dozen species of higher plants, including gingkos, cycads, alders (genus *Alnus*), buckthorns (*Caemothus*) and legumes such as peas, soybeans, alfalfa, clover and vetch (56, 153, 188).

Plants not associated with nitrogen-fixing organisms rely on the soil reserve of nitrogen compounds or on fertilizers. In ecosystems lacking symbionts, free-living bacteria often fix nitrogen at a rate of 2–6 kilograms per hectare per year (56). Nitrogen-fixing symbiotic organisms associated with legumes can easily fix 350 kilograms per hectare annually in crop fields (56). The importance of free-living nitrogen fixers should not be underestimated. In tropical rice paddies, non-symbiotic organisms have maintained the nitrogen supply for thousands of years without the aid of nitrogen fertilizers (281).

The mechanism of biological fixation is not fully understood. Basically, atmospheric nitrogen combines with hydrogen to form ammonia as follows:

$$N_2 + 160 \text{ kilocalories} \longrightarrow 2N$$
$$2N + 3H_2 \longrightarrow 2NH_3 + 13 \text{ kilocalories}$$

A net input of 147 kilocalories is required to fix one mole of nitrogen. The energy source for symbionts is the photosynthesis of the higher plants with which they associate. Free-living blue-green algae perform their own photosynthetic reactions.

Nitrogen-fixing bacteria may actually not require the amount of energy shown in the chemical equations, as they contain an enzyme called nitrogenase. Whereas industrial ammonia production requires extremely high temperatures and thousands of kilograms of pressure, nitrogenase enables the bacteria to accomplish the same task at ordinary pressures and temperatures (56). The total amount of nitrogenase in the world is probably no more than a few kilograms (56).

Future role of biological nitrogen fixation in world food production

The future role of biological nitrogen fixation in improving world food production is highly promising. The possibilities include:

(*a*) enhanced biological nitrogen fixation in leguminous crops;
(*b*) the extension of symbiotic nitrogen-fixing organisms to crops other than legumes:
(*c*) the transfer of genes controlling nitrogen fixation from one bacteria to another.

The worldwide production and yields of cereal grains have risen sharply since 1950 (Fig. 66) partly because of the increased application of nitrogenous fertilizers to the cereal crops. Legumes, being associated with nitrogen-fixing bacteria, show only a slight yield increase with nitrogen fertilizer application (108). It appears that improved biological nitrogen fixation may play a key role in increasing legume yields (Fig. 67).

The rate of photosynthate production in a higher plant limits nitrogen fixation by the symbiotic organisms of the plant. Figure 67 depicts the dramatic differences in nitrogen fixation at two rates of net photosynthate production by soybeans. Soybeans grown in carbon-dioxide-enriched air fixed 425 kilograms of nitrogen per hectare compared with only 75 kilograms for those grown in unenriched air (108). Respiration decreases as the carbon dioxide to oxygen ratio increases, thereby allowing a net increase in the production of photosynthate.

In 1973 about \$10 million were spent on nitrogen fixation research, which engaged about two hundred scientists (123). If sufficient break-

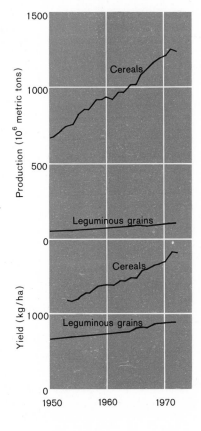

FIGURE 66. Comparison of world production and yields of cereals and leguminous grains. Since 1950, cereal yields have increased rapidly, primarily because of larger applications of nitrogen fertilizers; however, leguminous grain yields have shown little improvement because of the failure to develop an appropriate technology (110).

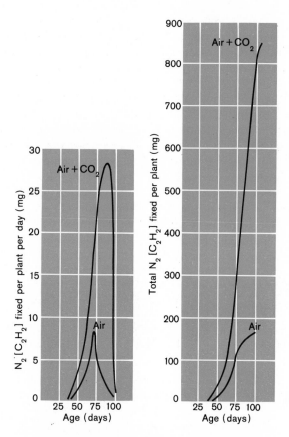

FIGURE 67. Daily nitrogen-fixing activity per plant and total amount of nitrogen fixed per plant for soybeans grown in carbon-dioxide-enriched air and in normal air. The total nitrogen input was increased from 295 to 510 kilograms per hectare (109).

throughs are made (and it appears there will be within the next ten years), their effect on agricultural productivity and energy consumption will be enormous.

Industrial processes require 422 to 654 $\times$ 10^8 joules for the manufacture of a metric ton of nitrogenous fertilizer (84). Besides, energy is expended for their transport and application. Nitrogen-fixing bacteria also require energy. In their symbiotic relationship, the plant supplies energy to the bacteria in exchange for the nitrogen fixed by the bacteria. As an estimated one sixth of the plant's energy may be used by the bacteria to form nitrogen, the result may be reduced crop yields.

Internal-combustion engines

To control crop growth, equipment is needed for soil preparation, irrigation, weed control and harvesting. The evolution of agricultural equipment has been greatly influenced by available power sources. Engine-powered agricultural equipment has improved in both speed and economy.

Engines are designed to exchange chemical energy for mechanical energy. Chemical combustion devices are classified according to whether or not the reactants and end products of the process pass through the mechanical conversion system. The reactants and end-product mixture of an internal-combustion engine work directly with the piston or turbine. About one third of the fuel energy becomes useful work, and the other two thirds dissipate as waste heat (112).

FUELS

Probably over 99% of the world's internal-combustion engines burn liquid fuel derived from petroleum. In some countries where natural petroleum is scarce, fuels of similar composition and characteristics are produced by coal hydrogenation.

Crude oil is the term used for raw petroleum as it comes from oil wells. It consists of a mixture of many different hydrocarbons of various molecular weights and includes a certain fraction — usually small — of organic compounds containing sulphur, nitrogen and other elements. The composition of crude oil differs widely according to its source.

Generally, crude oil falls into three classes depending on whether the residue from distillation is paraffin, asphalt or a mixture of the two. The corresponding crude oils are called paraffin-base, naphthenic-base or mixed-base.

Products of petroleum refinement are classified by their use and their specific gravity and volatility as determined by distillation at standard atmospheric pressure. The products of greatest interest are described in the following paragraphs.

Natural gas

Gaseous hydrocarbons are usually associated with liquid petroleum. They either occur above the liquid in the earth or are dissolved in it. The dissolved gas is the first product to separate out during distillation.

Petrol (gasolene)

The term covers most liquid petroleum fuels intended for use in spark-ignition engines. Petrol becomes 100% volatile between 162°C and 230°C. Its chemical composition varies widely, depending on the crude-oil base and refining methods.

Kerosene (vaporizing oil, lamp-oil)

This product is used in lamps, heaters, stoves and similar appliances. It is also an excellent fuel for compression-ignition engines and for aircraft gas turbines. Jet fuels, used primarily for aircraft turbines and jet engines, resemble kerosene in composition. The usual distillation range of kerosene is nearly 96% at 265°C.

Distillate

Although distillate is slightly heavier than kerosene, it has substantially the same uses. It is obtained from some western U.S. crude oils by distillation at atmospheric pressure. It becomes 98% evaporate at 285°C.

Diesel oils

These petroleum derivatives lie between kerosene and the lubricating oils. Their composition is controlled for use in various compression-ignition engines.

Fuel oils

The distillation and specific gravity range of fuel oils are similar to those of Diesel oils; but because they are designed for use in continuous burners, their composition does not need to be so accurately controlled as that of Diesel oils.

Lubricating oils

These are made up partly of heavy petroleum distillates and partly of residual oils from distillation (e.g., tar and asphalt).

SPARK-IGNITION (OTTO) ENGINE

An Otto engine operates on repetitious changes of the reactant gases (fuel and air) within the piston-cylinder system. The four-process cycle consists of (1) the intake stroke, (2) the compression stroke, (3) the power stroke and (4) the exhaust stroke.

The four-stroke engine is set in motion when the piston draws in the air-fuel mixture at constant pressure (intake stroke). Next, a compression stroke condenses the gases. When the piston motion stops, the mixture is ignited and burns rapidly. As the temperature of the gases increases, their expansion pushes the piston downward (power stroke). Finally, the expended gases are dispelled through the exhaust valve.

Spark-ignition engines operate on two different types of fuel: petrol (gasolene) and liquid petroleum (LP). Liquid petroleum is a mixture of butane and propane. Mixtures are usually preferable, because butane does not change from liquid to gas below freezing point, whereas propane gasifies at —42°C.

The energy capacity of butane is 29×10^6 joules per litre at 16°C and that of propane is 26×10^6 joules per litre, as compared with 38×10^6 joules per litre for petrol. LP gas has a high octane value, which increases compression ratios, causing the converted tractor to run smoother and apparently providing more lugging ability. However, an engine converted to LP gas may use as much as 20–25% more fuel than the same engine powered by petrol. Thus gasolene must be priced higher than LP gas.

Typical petrol (spark-ignition) engines have thermal efficiencies of 25–30%. Spark-ignition engines have been most useful in motor-cars, light trucks and wheel-type tractors. The earliest tractors had petrol engines, but their limited efficiency encouraged Diesel tractor development.

In 1970 there were an estimated 15.6 million tractors in the world (Table 61). The developed areas (Europe, U.S.A., Canada, New Zealand, Australia, South Africa and Japan) accounted for nearly 14.3 million tractors, and there were only about 1.4 million tractors in the rest of the world (15).

TABLE 61. – FOUR-WHEEL TRACTORS IN WORLD AGRICULTURE, 1970

	Arable land	Estimated number of four-wheel tractors	Power per tractor		Hect-ares per tractor	Power per hectare	
	10^3 hect-ares		Horse-power	Kilo-watts		Horse-power	Kilo-watts
Europe, U.S.A., Canada, Japan, New Zealand, Australia, South Africa	507 363	14 252 000	50	37	35.6	1.40	1.04
Rest of world .	924 637	1 364 000	40	30	678.0	0.06	0.04
TOTAL	1 432 000	15 616 000					

SOURCE: (269)

DIESEL ENGINE

On first appearance the Diesel engine may easily be mistaken for an Otto engine. In the Diesel engine, however, the gas is ignited by compression rather than by a spark.

In the compression-ignition engine usually a full, unthrottled charge of air (volume is constant) is drawn into the chamber during the intake stroke. The air volume decreases to one twelfth or one twentieth the original volume during the compression stroke. As the pressure changes less rapidly than the volume during compression, the air temperature increases to about 550°C. Just before the piston reaches top centre of the cylinder the fuel is sprayed into the combustion chamber, where it is ignited by the high air temperature and burns almost as soon as it is introduced. The combustion products are expanded and exhausted in the usual way (203).

Well-designed compression-ignition engines are usually more efficient than spark-ignition engines. Both spark-ignition and compression-ignition engines have proportionally greater efficiency with increasing compression ratios. With compression ratios of 14:1 to 22:1, Diesel efficiencies range from 30% to 36% (Table 62); in addition, Diesel engines gain efficiency even with partial loads.

HUMPHREY-PUMP ENGINE

The Humphrey-pump engine is a combined engine and pump unit intended for low-head applications. It operates much like the Otto engine, except that the metal piston within the engine block is replaced by a column of water.

The cycle begins with a compressed mixture of fuel and air above the water piston (Fig. 68). While the valves remain closed, a spark ignites the gaseous mix and drives the water down the cylinder and into the flowpipe where it continues under its own momentum. As the pressure in the cylinder drops below atmospheric pressure, the water valve opens and draws in water. The scavenge and exhaust valves open as the water returns (it oscillates in the pipework), expelling the exhaust gases. The exhaust valve closes as the returning water reaches its level and entraps a volume of compressed air, which acts as a cushion to force the water back down the cylinder. At this point the inlet valve opens, drawing a new mixture of fuel and air in through the carburettor. The water then returns, compressing the mixture until the pressure in the cylinder triggers another spark, and the cycle is repeated.

For better understanding of the operation of the Humphrey pump, tests are being made in Reading, U.K., using a model 15 centimetres in diameter. Fuelled by methane, it pumps about 400 litres per minute and has an overall efficiency of 10%.

Because of its simplicity, the Humphrey pump is of particular interest for manufacture and use in the developing countries. Once a reasonable prototype is developed, it will need testing under working conditions.

TABLE 62. – CHARACTERISTICS OF FOUR-STROKE COMPRESSION-IGNITION (DIESEL) AND SPARK-IGNITION ENGINES

Characteristics	Compression-ignition	Spark-ignition
Compression ratio	14–22:1	5–8:1
Ignition	Compression heat	Electric spark
Thermal efficiency	30–36%	25–30%
Fuel induction	Injector	Carburettor
Fuel used	Fuel oil	Usually gasolene
Fuel system	Fuel separate from air	Air-fuel mixture
Fire hazard	Less	Greater
Power variation	More fuel used	More mixture used
Air volume used	Constant	Variable
Air-fuel ratio	15–100:1	10–20:1
Relative fuel consumption	Lower	Higher
Energy per litre of fuel	Higher	Lower
Manifold throttle	Absent	Present
Exhaust gas temperature	482°C	704°C
Starting	Harder	Easier
Lubricants	Heavy-duty oil	Regular and premium oil
Speed range	Limited (600–2 500 rev/min)	Wide (400–6 000 rev/min)
Engine weight per horsepower	About 8 kg	Average about 4 kg
Initial cost	High	Much lower
Lugging ability	Excellent	Less

SOURCE: (106)

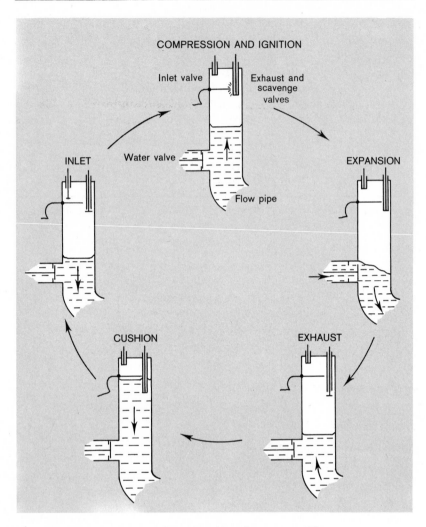

FIGURE 68. Principle of the four-stroke Humphrey pump (60).

GAS-TURBINE ENGINE

Gas-turbine engines, unlike Otto or Diesel engines, have no reciprocating parts. They are cylindrical with radial vanes, or paddles, along the outside perimeter. An energetic flow of fluid against the vanes turns the turbine. The power stroke is therefore a continuous rotating motion.

The main components of a single gas turbine are the compressor, the turbine and the combustion chamber (Fig. 69). Compressors both draw in fresh air and compress it. As pressurized air is directed into the combustion chamber, vaporized fuel is fed to a sparger. An abundant supply of fuel and air supports a "torch-like" continuous combustion process. Hot expanded gases are forced out of the chamber by pressure toward a ring nozzle alongside the turbine. Pressure decreases in the ring nozzle, causing higher gas velocities.

Now, a continuous power stroke exchanges gaseous kinetic energy for mechanical shaft energy. High-velocity gas directed against the turbine vanes pushes the wheel around on its fixed axis. A connected crankshaft rotates with the turbine to provide useful work.

A widely used device for electrical generation in recent years is the combined-cycle turbine. The primary generator is a gas turbine, and a waste-heat boiler is used as the secondary steam-turbine generator. The greatest advantages of this system are increases in both gas-turbine efficiency and the recycling of waste heat (10).

By 1980, gas turbines should be able to provide an output of one kilowatt-hour for an input of 9.5×10^6 joules (34% efficiency) by increasing the compression ratio from 10:1 to 16:1 without raising the turbine inlet temperature above 1 100°C (10).

From a practical standpoint, turbines fall into two classes: (1) dynamic, or aircraft, units; and (2) stationary, marine and railway power

FIGURE 69
Open-cycle gas turbine
(133).

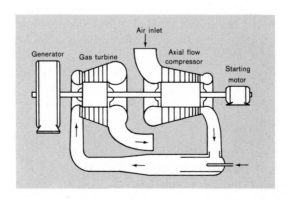

units (223). In jet and propeller turbine units for aircraft, thrust is produced either directly from the turbine (as in a jet) or indirectly from a rotating propeller. Stationary power units comprise generators, propellers and driving wheels. About 80% of the world's gas-turbine power in 1973 was used for electric power generation (83).

External-combustion engines

The primary distinction between internal- and external-combustion engines is that the combustion for operating the latter occurs outside the engine. A closed or open circuit of metal tubing containing a liquid or a gas passes through an external heated chamber. The vaporized liquid or the gas then performs useful work on the piston or turbine.

Heat is generated for the evaporation process by either chemical or nuclear energy conversion. A boiler, for instance, evaporates water and then superheats the steam through chemical fuel combustion.

Until about 1830, steam engines were less than 10% efficient. Today, steam turbines are about 40% efficient owing to the attainment of higher boiler pressures and temperatures (Fig. 70).

STEAM-PISTON ENGINE

The steam-piston engine, a traditional power source for railway locomotives, is one of the simpler double-acting engines. A piston and gliding valve operate from a main shaft as shown in Figure 71. The piston regulates the volume of the two chambers within the cylinder, while the gliding valve controls the direction of the steam flow to these chambers.

Operation begins when steam is forced into the piston chamber on the rod side. As the piston slides to expand the volume on the rod side, cool steam on the face of the piston is pushed out of the exhaust port. When the piston is fully extended, both valves are closed. The momentum of the shaft draws the gliding valve back to open another inlet port to the chamber on the face side. Hot steam surges against the piston and pushes it to increase the volume, thereby decreasing the volume on the rod side. Expansion forces the piston to extend fully again, but in the opposite direction, so as to exhaust the cooled steam. The cycle begins anew as the gliding valve slides to open the rod-side inlet and exhaust ports (242).

At the beginning of the twentieth century, steam-piston engines still produced more than 90% of the world's mechanical power. By 1976 this share had fallen to 10% or less owing to the predominance

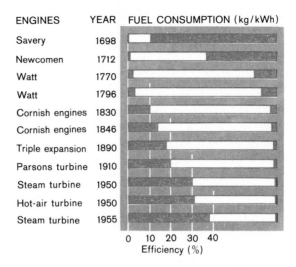

ENGINES	YEAR	FUEL CONSUMPTION (kg/kWh)
Savery	1698	
Newcomen	1712	
Watt	1770	
Watt	1796	
Cornish engines	1830	
Cornish engines	1846	
Triple expansion	1890	
Parsons turbine	1910	
Steam turbine	1950	
Hot-air turbine	1950	
Steam turbine	1955	

0 10 20 30 40
Efficiency (%)

FIGURE 70.   Progress in steam-engine efficiency (242).

of internal-combustion engines in small units of up to a few hundred kilowatts and of turbines in larger units of 7 500 kilowatts and more. The use of steam-piston engines is now restricted mostly to railway locomotives, smaller steamboats, and special engines designed for relatively slow running speeds (242).

STEAM TURBINE

The steam turbine is similar in principle and operation to the gas turbine. Steam is generated externally through combustion or a

FIGURE 71
Double-acting piston of a steam engine (242).

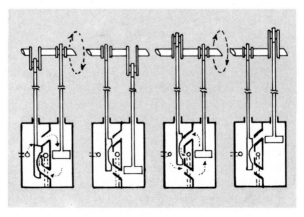

nuclear reaction. High-pressure vapour circulates to a ring nozzle, where volume increases, pressure decreases and velocity increases. Consequently, the kinetic energy of the high-velocity vapour is converted into mechanical energy as the turbine and turbine shaft rotate. The exhausted vapour recirculates to the boiler chamber, sometimes passing through other steam turbines or heat exchangers.

Overall efficiency is the product of three components: (1) boiler efficiency, (2) thermodynamic efficiency and (3) blade efficiency. In the boiler the burning chemical fuel heats the metal boiler surface. Convection to the water in the circulating tubes causes an increase in water temperature and potential energy. The ratio of the energy in the vapour to the chemical energy used is the boiler efficiency.

Next, the condensed vapour expands within the ring nozzle over the turbine. The ratio of the kinetic energy within the gas as it leaves the nozzle to the potential energy of compression determines the thermodynamic efficiency.

Finally, the turbine vanes intercept the gas, thereby turning the turbine shaft. The ratio of the mechanical energy of the turbine to the kinetic energy of the gas is the blade efficiency.

Efficient operation depends on all three of these factors. Although individual steam turbines operate at about 750°C with 41% efficiency, a complete plant, including the boiler, realizes only 34% efficiency.

Steam turbines are superior, however, to steam-piston engines. Turbines operate at higher speeds with less weight and volume. The power conversion is effected by a smooth, vibrationless rotary motion. Most important, efficiencies are higher as shown in Figure 70.

Economic efficiencies are obtained with more powerful steam turbines designed to deliver 750 to 750 000 kilowatts. At these ratings, steam turbines are stationary units for electric power generation. Distribution curves for steam engines of different types in the U.S.A. by size and total output are shown for 1870 and 1950 in Figure 72.

STIRLING ENGINE

The Stirling engine works by alternately heating and cooling a gas in a confined space. The rising and falling gas pressure causes a separate output piston to move back and forth in a cylinder to perform useful work by, for example, pumping or turning a wheel (Fig. 73).

Heating and cooling are carried out by pushing the gas to and fro between a section of the engine that is kept hot by an external heater and another part that remains cool. In Stirling's original design, as well as in most later versions, this pushing is done by a crank mechanism driven by the power output piston. Because the gas has only

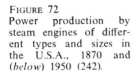

FIGURE 72
Power production by steam engines of different types and sizes in the U.S.A., 1870 and (*below*) 1950 (242).

A = Piston steam engines
B = Internal-combustion engines
C = Steam turbines

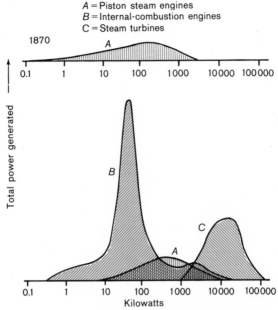

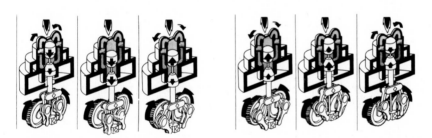

FIGURE 73. Principle of the Stirling engine (282).

to be moved, not compressed, this part of the action requires very little effort (13).

Such engines have had little success because of the difficulty of designing an efficient high-temperature regenerator of reasonable size. Recent developments in high-temperature alloys and heat transfer processes have led to new attempts to improve the Stirling engine.

The efficiency range of the Stirling engine is competitive with Diesel and gas-turbine engines. The early Stirling engine was used to power butter churns and grinders, but it met with only moderate success and quickly faded as a prospective power unit. Nevertheless, the Phillips Company in the Netherlands has taken the lead in reviving interest in the Stirling engine. The General Motors Corporation in the U.S.A. also conducted considerable work on the engine between 1958 and 1968. In 1968 further agreements for developmental work were made with two German companies, MAN of Augsburg and MWM of Mannheim. Moreover, the Swedish Consortium of United Stirling and many universities throughout the world are working to make it a competitive engine. Uses under investigation include power for satellite space stations, urban buses, portable generating sets, marine engines and small power generators for the developing countries.

Electric energy

Electricity plays an important role in stimulating economic growth and social well-being. In highly industrialized countries the entire social and economic structure depends on it, although electricity expenditures amount to only about 1 % of the gross national income (21).

Electricity has been available to the public in some highly industrialized countries for nearly a century, but there is no sign of a "saturation" of demand, although annual growth rates tend to decline slowly with time. Even in countries where the yearly consumption of electricity is thousands of kilowatt-hours per person, consumption is still increasing by 5–7 % annually. Although this rate may decline gradually, the annual increase in consumption is arithmetically greater each succeeding year.

Yet, there are vast areas of the world where the benefits of electricity are unknown to hundreds of millions of people. Their economic and social development would be greatly accelerated if electricity were to be made available to them. Electricity is the foremost potential stimulus to industry (even small-scale cottage industries) because its effect on production is far greater than its cost. Electricity is likewise a primary means of bringing comfort and health.

The electric system

Most electric energy is produced by mechanically driven generators. Both direct and alternating current may be generated by passing coils of wire through a magnetic field.

The principles of the direct-current (DC) generator are illustrated in Figure 74. A simple horseshoe magnet forms a magnetic field as magnetic "lines of force" travel from one pole to the other. A loop

FIGURE 74. Principle of a simple direct-current generator (182).

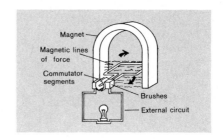

of wire supported on a shaft rotates within this field. As the loop turns, an electric current is generated in it and flows to terminals (segments) on the commutator. Two stationary contacts, or brushes, touch the commutator terminals as the loop of wire rotates. The current generated in the loop is carried by the brushes to an external circuit, where it is used as electric energy.

The commutator consists of two segments, so that with each half turn of the loop each brush makes contact with the other half of the commutator. This keeps the current flowing in the same direction in the circuit, thereby generating direct current.

Direct current is used primarily in mobile equipment and special industrial applications. A serious disadvantage of direct current is the difficulty of transmitting over long distances (182).

An alternating-current (AC) generator (often called an "alternator") operates on much the same principles as the DC generator, as can be seen in Figure 75. It also has a wire loop in which an electric current is generated by the rotation of the loop through the lines of force in a magnetic field.

However, the ends of the loop, instead of being connected to segments on a commutator as in the DC generator, are here connected to "slip" rings, which are fastened to the shaft and rotate with it. Stationary brushes are in contact with the slip rings. Current is carried through the brushes to an external circuit, which conducts the electric energy to the place of use.

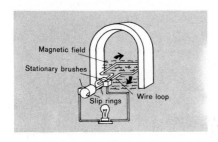

FIGURE 75. Principle of a simple alternating-current generator (182).

During one revolution of the loop each side passes through the magnetic lines of force, first in one direction and then in the other. The current thus produced is called alternating current because it flows first in one direction and then in the other (182).

How MECHANICAL GENERATORS ARE DRIVEN

Water power

Water flowing from a higher to a lower level can be used to turn a generator. This action is called hydroelectric generation. Large streams are dammed to create reservoirs to provide a controlled water flow (Fig. 76).

The principle of the waterwheel is used to turn the hydroelectric generator (Fig. 77). The "waterwheel," made of metal, is called a turbine wheel (Fig. 78). The generator is attached to the turbine shaft. Water directed against the turbine blades thus turns both the turbine and the generator (182). Hydropower is discussed in detail on pages 225–229.

Steam power

Water is heated in a boiler until it vaporizes. The steam at high temperatures and pressures is directed against the blades, or fins, of

FIGURE 76. Hydroelectric generating plant (182).

FIGURE 77. Waterwheel (182).

FIGURE 78. Principle of a water-powered generator (182).

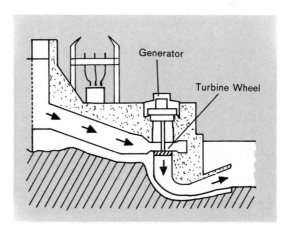

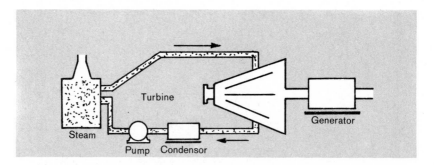

FIGURE 79. Principle of a steam-powered generator (127).

the turbine, causing it to turn. The generator, being connected to the turbine, also turns (Fig. 79).

Heat to produce steam is generated from the following sources:

1. *Fossil fuel:*

   (a) *Coal* is the most commonly used fuel for making steam-generated electric energy. Coal is still rather abundant in many regions of the world and can be mined at relatively low cost. Massive pulverizers reduce the coal to dust, which is blown into the boiler's firebox under high pressure. It burns almost instantly, creating enough heat to produce steam (Fig. 80).

   (b *Oil* — mostly of the heavy residual type — is also used for fuel in steam generating plants (Fig. 81).

   (c) *Natural gas* is used in some steam generating plants.

2. *Nuclear power.* Scores of nuclear plants operate with the same types of steam turbines used with other fuels. Nuclear power is discussed in detail on pages 177–183.

3. *Geothermal energy.* Geothermal energy is limited to certain areas. It is produced by the seepage of water through normal seismic faults to heated rocks many metres below the surface. There it is converted to steam, which can be used to operate generators. Natural geysers are potential sites for geothermal electric power plants (182). Geothermal energy is discussed further on pages 206–211.

4. *Solid wastes.* The use of municipal solid wastes to fuel steam-powered generators contributes to the disposal and recycling of such materials. At present, only a small portion of the electric energy requirements is furnished by solid-waste power plants (182).

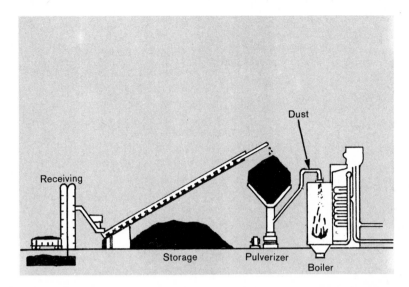

FIGURE 80. How steam is produced in a coal-burning generating plant (182).

FIGURE 81. How steam is produced in an oil-burning generating plant (182).

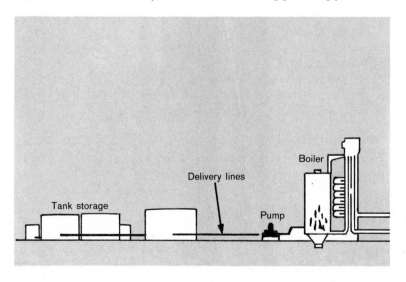

Internal-combustion engines

Small-scale electrical generators are sometimes powered by internal-combustion engines of three types: turbines, Diesels and gasolene engines.

ENERGY SOURCES

Coal and petroleum products are the most common fuels for the generation of electricity. Where possible, local energy sources are preferred over imported ones. This is not always feasible, but the advantages of relying on indigenous resources and curtailing imports are sufficient to justify serious examination of local reserves.

In remote areas the high cost of transporting oil may favour the use of reciprocating steam engines that can be fuelled with bagasse or other waste vegetable products. If there is a waterfall within reasonable distance of the centre of consumption, its possible use should be investigated and compared with alternative methods of generation. Sometimes wind power may offer the only practical solution. If the choice is close, only careful production cost estimates and common sense can determine the better method.

HOW ELECTRIC ENERGY IS TRANSMITTED

Electric energy is unique because it must be used the instant it is generated. It cannot be stored economically. This creates many problems for power suppliers who must generate electric energy to meet varying demands.

As electric energy is generated, it is transformed and transported instantaneously to the consumer through a network of wires known as transmission and distribution lines (Fig. 82). To avoid excessive losses during transmission, voltage and amperage are altered by transformers. A group of transformers is called a substation.

Step-up transformers

Large amounts of electrical energy (thousands of volts and amperes) are generated at the power plant. Because of resistance in the wires to the flow of electrical energy, some of the power is lost during transmission. Lowering the amperage also reduces the power loss in transmission and distribution lines.

Step-up transformers are used at the power plant to increase voltage and decrease amperage (Fig. 83). Consequently, efficiency is increased and smaller transmission wires can be used. In the step-up trans-

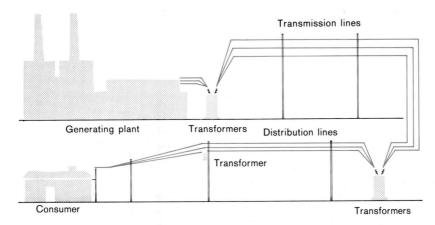

Transmission lines

Generating plant   Transformers   Distribution lines

Transformer

Consumer   Transformers

FIGURE 82. How electric energy is generated, transformed, transmitted and distributed to the consumer (182).

former the power source is connected to the coil with the least number of turns. Electric energy is induced in a second coil. Power taken from the second coil decreases the amperage in proportion to the increase in voltage. This change in voltage and amperage is proportionate to the number of turns in each coil.

Step-down transformers

Before electric power can be used, the voltage is stepped down and the amperage is stepped up. This is done first when the power is transferred from transmission to distribution lines and again when it is transferred from distribution to service lines (Fig. 84).

In the step-down transformer the power source is connected to the coil with the greatest number of turns. Power taken from the coil with the fewest turns decreases voltage and increases amperage.

FIGURE 83. Principle of the step-up transformer (182).

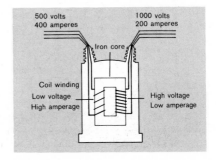

500 volts
400 amperes

1000 volts
200 amperes

Iron core

Coil winding
Low voltage
High amperage

High voltage
Low amperage

FIGURE 84. Principle of the step-down transformer (182).

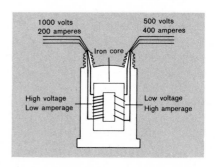

Transmission lines are used to move large quantities of electric energy from step-up transformers at the generating plant to the distribution area. Voltages range from 36 000 to 750 000 volts. The distance may be several hundred kilometres. These lines are usually supported on steel or aluminium towers over 30 metres high (Fig. 85).

Electrical energy is delivered to users through lines that fan out over the countryside from step-down distribution substations. Distribution lines may be suspended overhead on poles or placed underground. Underground systems do not mar the landscape and are better protected from weather hazards.

FIGURE 85. Transmission lines for transporting electric energy from the generating plant to the distribution substation (182).

RURAL ELECTRIFICATION IN DEVELOPING COUNTRIES

Developing countries are allocating increasing resources to rural electrification (Table 63) for the purpose of furthering economic and social aims (248). The primary objective is to improve the standard of living in the rural areas and to prevent an influx of population in urban areas by the industrialization of rural areas (152). Most countries are still in need of further rural electrification, but the possibilities and type vary by country.

Technology and costs

Rural areas receive electricity from (1) autogenerators serving single consumers, (2) autogenerators serving several consumers on a local network and (3) the main grid system of the utility company.

Autogenerators are powered by Diesel engines or small turbines. They range in size from about 5 kilowatts (sufficient to meet such minor needs as refrigeration and lighting on a farm) to over 1 500 kilowatts (sufficient to meet the power needs of a large sugar processing plant).

TABLE 63. – EXTENT OF RURAL ELECTRIFICATION BY WORLD REGIONS, 1971

| | Population | | | Village-rural population served [2] | |
| | Total | Village-rural [1] | | | |
	..... Millions		Percent	Millions	Percent
Latin America	282	140	50	32	23
Selected countries in Europe, the Near East and North Africa [3]	143	87	61	45	15
Asia	934	700	75	105	15
Africa	182	165	91	7	4
	1 541	1 092	71	189	12

SOURCE: (127)

[1] The definitions of "village" and "rural" vary. Generally, villages are conglomerations of 5 000 to 10 000 inhabitants or less, whereas rural refers to low-density population outside villages, often living in clusters close to large farms. – [2] Electrification data are not available for every country; so the percentages should be taken as typical levels for countries in the region, although there may be considerable variance. – [3] Algeria, Cyprus, Egypt, Iran, Morocco, Saudi Arabia, Tunisia and Turkey.

The utility company often decides whether to provide electricity from the grid or from local autogenerators. This decision depends on numerous factors, including expected level and growth of demand, expected use of the investment, distance from the main network and difficulty of the terrain. Table 64 lists typical cost data at two levels of demand.

The capital costs of grid supplies are much higher than those of autogeneration, but fuel, operation and maintenance costs are much lower. High use favours the more capital-intensive investment and lower fuel costs for supplies from a grid. The relative annual costs of autogeneration and supplies from a grid for 50-kilowatt projects at various use levels are given in Table 65.

The high fuel bill is a serious drawback of autogeneration. Generally, autogeneration compares well with public supplies from a grid only at low use levels or when demand is widely scattered. Extending a subtransmission link 25 kilometres to an isolated demand centre could cost about $100 000 ($10 000 per year at 10% annuity). Table 66 shows the effect of transmission distance from the grid on capital costs of public supplies (50-kilowatt capacity). Table 67 shows the effect of distance and load factor on the cost comparisons in Table 66.

TABLE 64. – TYPICAL COSTS OF PUBLIC SUPPLIES AND AUTOGENERATION OF ELECTRICITY (1972 DATA)

	Autogeneration		Supplies from grid [1]	
Capacity of project (kilowatts)	50	25	50	25
Consumers served	140	70	140	70
Capital costs	$34 000	$25 000	$56 000	$38 000
Fuel, operation, and maintenance costs per kilowatt-hour	$0.06	$0.06	$0.005	$0.005
Annual billing, administration, and other costs	$2 000	$1 000	$2 000	$1 000

SOURCE: (127)

[1] Average length of subtransmission line per village is taken as 4 kilometres. Note the economies of scale in capital costs. The 50-kilowatt and 25-kilowatt projects could fully serve villages of about 2 000 and 1 000 inhabitants, respectively. Farms and agro-industries outside the village might add from 20 kilowatts to 1 000 kilowatts or more to total demand. Capital costs range from $400 to $550 per consumer for supplies from the grid (or $40 to $55 per caput in the village served); however, for larger villages of 5 000 to 10 000 inhabitants, these costs might drop to $200 per consumer ($20 per caput) or less.

TABLE 65. – RELATIVE ANNUAL COSTS OF 50-KILOWATT PROJECTS BY LOAD FACTOR

	Autogeneration			Supplies from grid		
Load factor	*10%*	*25%*	*50%*	*10%*	*25%*	*50%*
Annual capital costs	$4 500	$4 500	$4 500	$5 600	$5 600	$5 600
Fuel, operation and mainte- nance	2 600	6 600	13 200	200	500	1 000
Billing, adminis- tration, etc. ..	2 000	2 000	2 000	2 000	2 000	2 000
Total	$9 100	$13 100	$19 700	$7 800	$8 100	$8 600
Average per kilo- watt-hour	$0.21	$0.12	$0.09	$0.18	$0.07	$0.04

SOURCE: (127)

TABLE 66. – EFFECT OF TRANSMISSION DISTANCE FROM GRID ON CAPITAL COSTS

Costs \ Distance from grid	4 kilo- metres	29 kilo- metres
Generation and transmission costs	$24 000	$ 24 000
Subtransmission costs	18 000	118 000
Local distribution .	14 000	14 000
Total	$56 000	$156 000
Annual capital costs	$ 5 600	$ 15 600

SOURCE: (127)

TABLE 67. – EFFECT OF LOAD FACTOR ON AVERAGE COSTS (CENTS PER KILOWATT-HOUR)

Load factor (%)	Supplies from grid		Autogen- eration
	4 kilo- metres	29 kilo- metres	
10	18	40	21
25	7	17	12
50	4	8	9

SOURCE: (127)

NOTE: As average costs in urban areas are about 3 cents per kilowatt-hour, the average costs for subtransmission networks to meet small demands in areas remote from the grid are obviously inordinate. However, the same subtransmission networks can be used to meet much larger demands. If adequate demand develops from farms, agro-industries and several villages, average costs decline very quickly to about 4 cents to 8 cents per kilowatt-hour.

Rural electrification projects pass through four phases:

1. Only a few scattered businesses need and can afford electricity.
2. A small collective demand develops and is met by a local network fed from autogenerators.
3. The collective demand becomes large enough to justify a grid system.
4. Low demand centres near networks are connected to them at low marginal cost.

The relative magnitude of these phases can be gauged from Table 63 and from the data for Mexico in Table 68, which shows that over 50% of the rural population there lives in areas of low demand. Similar patterns of village size and demand exist in most countries. (In Cameroon and Tanzania, emphasis is placed on the formation of large villages before providing electricity and other infrastructure.)

Uses of electricity

Electricity is used in rural areas for a wide variety of household and "productive" purposes (247). Generally, the demand stemming from "productive" uses is higher than that from household uses (Table 69). Consumption levels in rural areas are much lower than in urban areas, but demand often grows considerably once an area is electrified (Table 70).

TABLE 68. – POPULATION DISTRIBUTION OF VILLAGES AND EXTENT OF ELECTRIFICATION IN MEXICO

Population ranges	Number of villages	Population		Remarks
		Number	Percent	
Less than 100 ..	55 376	1 823 900	7	Low-demand areas, 25% electrified
100–499	28 494	6 944 500	26	
500–999	7 346	5 091 900	19	
1 000–4 999	5 207	9 681 800	37	Medium- to high-demand areas, 30% electrified
5 000–9 999	416	2 894 300	11	
	96 839	26 436 400	100	

SOURCE: (41)

TABLE 69. – PERCENTAGE DISTRIBUTION OF THE DEMAND FOR ELECTRICITY IN URBAN AND RURAL AREAS OF SELECTED COUNTRIES (1971 DATA)

| Country | Rural | | | | | Urban | |
| | "Productive" | | | | House-hold | "Pro-ductive" | House-hold |
	Farms	Agro-industry	Com-mercial-com-munity	Total			
Chile	9	26	32	67	33	...	...
China	10	16	16	26	74	80	20
Costa Rica	—	—	—	70	30	43	57
El Salvador	—	—	—	45	55	60	40
Ethiopia	—	—	—	55	45	44	56
India	59	21	21	80	20	89	11
Nicaragua	15	45	45	60	40	30	70
Pakistan	23	17	17	40	60	90	10
Tanzania	—	—	—	75	25	80	20

SOURCE: (127)

TABLE 70. – ESTIMATED LEVEL AND GROWTH OF CONSUMER DEMAND FOR ELECTRICITY IN URBAN AND RURAL AREAS OF SELECTED COUNTRIES (1971 DATA)

| Countries | Demand per consumer [1] | | Annual growth rate of demand | |
	Rural	Urban	Rural	Urban
	.. Kilowatt-hours per year ..		.... Percent	
Costa Rica	1 900	6 000	20	10
El Salvador	1 000	4 000	20	10
Ethiopia	800	2 000	40	15
India	1 000	...	15	10
Thailand	200	4 000	12–20	22

SOURCE: (127)

[1] Rural data are for selected areas; urban data are for capital cities, except for Thailand, which is an average.

Consumer demand varies between areas, depending on the types of "productive" uses. Irrigation pump sets, for example, consume about 3 000 kilowatt-hours per year in India, while a large agro-industry may consume 100 000 kilowatt-hours or more annually.

Aims and conflicts

Apart from being far superior to the alternatives, electricity increases energy use and often reduces energy costs in the area; but this is generally not the case during the first years of implementation for two reasons:

1. Initial investment costs are high for low-density populations, often remote from main networks.
2. Over ten years may pass before demand develops to full network capacity.

Rates are often kept low in relation to costs so as to provide cheap energy to low-income households and small businesses during the first years (Table 71).

Effects of oil price increases

Rising oil prices have greatly increased the costs of electricity from Diesel-powered autogenerators (roughly 50% to 100% depending on use), of kerosene lighting and of Diesel motive power used in irrigation

TABLE 71. – COMPARATIVE ELECTRIC POWER DATA FOR URBAN AND RURAL AREAS IN EL SALVADOR

	Urban	Rural
Annual use per consumer	4 000 kWh	600 kWh[1]
Load factor	50%	20%[1]
Investment in subtransmission and distribution per consumer (approximate)	$100	$300
Average costs per kilowatt-hour	$0.025	$0.06–0.10[1]
Average price per kilowatt-hour	$0.028	$0.04
Mean per caput incomes	$800	$125

SOURCE: (127)

[1] Figures refer to typical initial conditions.

and agro-industries (30 % to 60 % depending on use). The effect should be an increase in the number of households and businesses using electricity. The cost of electric power in areas to be electrified will depend on the mix of hydro, coal and oil plants in the system. The consumer response data in this report are from before the oil price increases; 1972 cost and price data are also used.

In African countries there are programmes for the electrification of small towns and the larger villages and businesses near them. Autogeneration is the main option, although public supplies can be contemplated in areas close to main networks. However, in ten years these programmes are unlikely to be serving more than 10 % of the rural village population.

Several Asian countries are striving for electrification of larger and medium-sized villages and the surrounding farms and agro-industries. In some cases, as in parts of India and China, the electrification programmes (based on public supplies from grids) also extend to smaller villages. In these areas, solar energy and small hydro plants may be suitable alternatives to rural electrification (166). Smaller hydro plants have been used to supply isolated consumers with a very low demand for electricity in Indonesia (152). Autogeneration is an alternative for meeting small or remote demands. About one fourth of the rural village populations in the Asian countries with electrification programmes may receive service within ten years.

In Latin America, several countries are planning to electrify the larger villages and surrounding areas, while others have completed this phase and are extending electrification to smaller villages and to new farm and agro-industrial consumers. Again, public supplies should replace local autogeneration, which is becoming extinct except in remote or low demand areas. About one third or more of the rural village populations in the Latin American countries may be served within ten years (127).

THE AFRICAN SITUATION

The development of electric energy has recently surged forward in Africa. Table 72 summarizes the situation in Africa in 1973, and Table 73 shows the production of electricity by type and sector in the same year. The nine countries listed in Table 74 produced 99×10^9 kilowatt-hours, almost 89% of the total output of the continent, in 1973. South Africa accounted for nearly 58% of this production, and

TABLE 72. – ELECTRICITY IN AFRICA, 1973

Total electricity production	111 831 GWh
Total installed capacity	28 904 MW
Use of installed capacity	3 869 h
Per caput consumption	294 kWh

SOURCE: (252)

NOTE: 1 kWh = 1 000 watt-hours; 1 GWh = 10^6 kilowatt-hours; 1 MW = 10^6 watts.

TABLE 73. – TOTAL PRODUCTION OF ELECTRICITY IN AFRICA BY TYPE AND SECTOR, 1973

	Public sector	Industrial sector
	.... 10^6 kilowatt-hours	
Thermal	72 996	7 464
Hydro	27 003	4 368
Total	99 999	11 832

SOURCE: (252)

TABLE 74. – LEADING AFRICAN PRODUCERS OF ELECTRICITY, 1973

Rank	Country	Production (10^6 kWh)	Percent of total production in Africa
1	South Africa	64 857	58.00
2	Egypt	8 104	7.25
3	Rhodesia	7 277	6.51
4	Zaire	3 884	3.47
5	Ghana	3 600	3.22
6	Zambia	3 419	3.05
7	Algeria	3 000	2.68
8	Morocco	2 639	2.36
9	Nigeria	2 625	2.35
	Total	99 405	88.89

SOURCE: (249)

the eight others — all developing countries — listed in the table for 31%, or 34 × 10⁹ kilowatt-hours. This represents almost 74% of the electricity produced by African developing countries in 1973. These are the only countries that produced over 2 000 million kilowatt-hours each. The only other African countries producing over 1 000 million kilowatt-hours were Cameroon (1.15 × 10⁹ kWh) and Tunisia (1.13 × 10⁹ kWh).

At the end of 1973 the total production of electricity in the African countries amounted to only about 7% of the continent's hydroelectric potential. Perhaps because of the time required for studying and harnessing water resources, African countries often resort to small, medium or large thermal installations to meet their urgent needs even if favourable hydroelectric opportunities exist close to the centres to be served. A wide range of types and sizes of generators is used. Low-powered Diesel generating sets in blocks of 50 to 1 000 kilowatts and consuming gas oil have been used to supply isolated centres. For capacities between 1 000 and 10 000 kilowatts, slow Diesel engines, Diesel gas generators or "twin" units (two engines driving a single alternator) have been used successfully.

The internal-combustion engine

The internal-combustion engine has proved satisfactory for electric power generation in Africa. Its use will be justified for a long time to come — especially in countries with few exploitable water resources — because of (1) the difficulty of providing an interconnected network that will serve the most distant regions and (2) the small demand for power in some isolated centres.

Steam-generated power

Steam-generating equipment is often used in Africa for power blocks of a few to several hundred thousand kilowatts. It is economical and safe to operate as well as durable. In coastal ports and towns, steam is produced by burning fuel oil.

Natural gas

The use of natural gas has completely altered electric energy production in Algeria, Tunisia, Libya, Nigeria and Egypt — all of which are countries with substantial gas reserves. Between 1970 and 1974, Algeria installed a capacity of 600 megawatts: 450 from conventional thermal stations and 150 from gas turbines. The country now has major stations at Skikda (270 megawatts), Oran (189 megawatts), Annaba (184 megawatts, due to receive an additional 130) and the port of Algiers (120 megawatts).

Coal

Coal-burning thermal stations are most important, especially in South Africa. That country has the most powerful thermal stations in the Southern Hemisphere: Camden (1 600 megawatts, equipped with eight units of 200-megawatt capacity), Arnot (capacity to rise to 2 100 megawatts, with units of 350-megawatt capacity) and Hendrina (2 000-megawatt capacity planned). South Africa also has other less powerful thermal stations. In Morocco the Jereda station has a capacity of 165 megawatts.

Geothermal energy

In the future, Africa can count on good supplies of geothermal energy, especially along the Rift Valley, in the east, with its many active volcanoes and hot springs. Despite the geothermal potential from the Red Sea to Lake Malawi (or Nyasa), Africa has only one low-capacity (275-kilowatt) geothermal station at Kwabukwa, in Shaba province, Zaire, which produces power to operate a tin mine about 10 kilometres from the station (24).

Nuclear energy

In Africa there are only a few small research installations; however, several projects are being studied in South Africa and Egypt.

An 800-megawatt nuclear station, built and operated by the Koeberg Power Station north of Cape Town, should be in service by 1981. South Africa's Atomic Energy Board is working toward setting up other nuclear stations before the end of the century to ensure the country's self-sufficiency in energy as the major African producer of uranium.

Large nuclear reactors do not appear to be justified in African developing countries because of their low consumption of electricity and the availability of more economical and suitable resources for their level of economic development. (Egypt and Libya are possible exceptions.)

Transmission of electric power in Africa

Electricity can be transmitted with relatively low energy losses over a few hundred kilometres, but beyond about a thousand kilometres the loss is substantial.

Before the end of the 1970s the Cabora Bassa hydroelectric station in Mozambique will be linked by 535-kilovolt lines to Pretoria and Johannesburg, the major centres of consumption in South Africa, over a distance of about 1 400 kilometres.

Construction of the transmission line from the Inga hydroelectric station to Kolwezi, in Shaba province, Zaire, was completed in 1976. This dual-circuit 500-kilovolt DC line covers about 1 800 kilometres and involves the most advanced technology of semiconductors.

Imaginative solutions will be needed during the next decades owing to the rapid expansion of electric power consumption and the depletion or exhaustion of some resources. The minimum projections show that consumption in Africa will approach 200 thousand million kilowatt-hours in 1980. Interconnections will become essential as African countries seek to promote effective multinational cooperation in the field of electric energy.

Rural electrification in Africa

Progress in rural electrification has been slight in all but a few countries, such as Algeria, Ivory Coast and Nigeria. Some have endeavoured to enhance electrification by maximum network extension in the countryside; others prefer many low-power autogeneration stations and independent distribution networks.

The rural electrification problem is complex, especially for countries with limited capital, in a continent as vast and sparsely populated as Africa. Areas to be supplied are often far and widely separated from one another, have inadequate transportation and communication systems, are sparsely populated, lack skilled manpower, maintain low-level economic activity and have unfavourable climates.

Because the density of the equipment diminishes with distance from the major towns, location of generating stations in or near larger centres is almost obligatory.

In rural areas, electrification will bring energy to smaller settlements for pumping water, lighting and domestic uses, as well as for the development of craft industries. It will make possible the setting up of small processing or manufacturing industries, thereby raising both the standard of living and the level of skills of the inhabitants. Diversification of employment and occupations can contribute to suitable training of local manpower for the immediate requirements of development.

Nuclear energy

Conventional steam-electric power plants burn fossil fuels (coal, oil or natural gas) in a boiler and use the resulting heat to generate steam. The steam drives a turbogenerator, thereby producing electricity. In a nuclear power plant the heat is furnished by a nuclear reactor (Fig. 86).

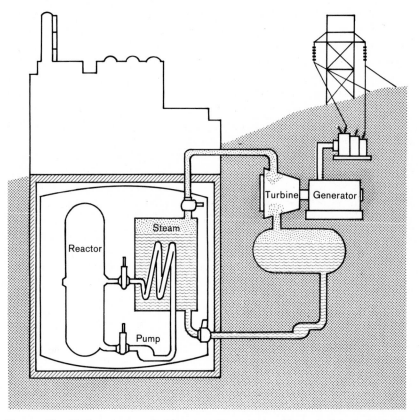

FIGURE 86. Nuclear-reactor electric generating plant. The principle is basically the same as that of the conventional fossil-fuel plant except for the fuel-burning system (260).

FISSION REACTORS

All commercial nuclear plants are based on fission. In this process the nucleus of an atom is bombarded by free neutrons, causing the nucleus to split and release its own neutrons, which, in turn, strike other nuclei. This chain reaction releases energy in the form of heat and gamma radiation.

The reactor vessel consists of a core containing a fuel, either uranium or thorium. A working fluid is pumped through the vessel and heated as it passes upward through the nuclear fuel. The type of fluid (usually boiling water, pressurized water, helium, carbon dioxide or heavy water) depends on the reactor. The heated fluid produces steam, which then passes through the turbine. The shaft turns the generator, and produces electric power (150). The commercial reactors in operation as of 1974 are listed by country in Table 75 (176).

The typical reactor produces 3 300 megawatts of thermal energy (MWt) and 1 000 megawatts of electrical energy (MWe) (197). Plant efficiency ranges from 33% to 40%, whereas a modern coal-, gas- or oil-fuelled power plant is about 40% efficient (33). Capital costs for coal plants can be 40% lower than those for nuclear plants — about $450 per kilowatt for coal versus $700 per kilowatt for nuclear plants (117); however, nuclear units have a fuel cost advantage over coal units. The capital cost advantage of coal plants is less than half the fuel cost advantage of nuclear plants. Moreover, the net energy yield for nuclear power plants is about four times the total energy input (69).

A nuclear reactor fuelled by half a kilogram of uranium supplies as much heat as a coal-burning plant using 22 000 kilograms of coal (194). It is estimated, however, that the uranium needed to fuel existing and planned non-breeder ("burner") nuclear power plants will deplete

TABLE 75. – COUNTRIES WITH NUCLEAR POWER PLANTS IN OPERATION, UNDER CONSTRUCTION AND IN PLANNING STAGE, 1974

Countries with plants in operation				Countries with plants under construction or in planning stage	
	No.		*No.*	Austria	Mexico
U.S.A.	50	Switzerland ..	3	Brazil	Philippines
United Kingdom ..	29	German Dem. Rep. .	2	Chile	Portugal
U.S.S.R.	16	Netherlands .	2	China (Taiwan province)	Singapore
France	10	Argentina ...	1	Denmark	South Africa
Germany, Fed. Rep. of	8	Belgium	1	Egypt	Thailand
Japan	7	Bulgaria	1	Finland	Turkey
Canada	7	Czechoslovakia	1	Greece	Yugoslavia
India	3	Pakistan	1	Iran	Hungary
Italy	3	Sweden	1	Israel	Poland
Spain	3			Jamaica	Romania
		TOTAL	149	Korea, Rep. of	

SOURCE: (176)

supplies within twenty years (33). A possible solution is the breeder reactor.

BREEDER REACTORS

Theoretically, more energy is stored during fission in a breeder reactor than is taken out in the form of heat. Only the fissionable uranium-235 isotope, which constitutes less than 1% of the earth's uranium, can be used in the conventional reactor described previously. The remaining 99% occurs as the uranium-238 isotope. Its nucleus contains more neutrons than that of the U-235 isotope and, consequently, does not split under neutron bombardment; but it can be converted into a fissionable material by capturing neutrons.

When fissionable material in the core of the breeder reactor is split by free neutrons, two or three neutrons are released from each nucleus. Some continue the chain reaction; others are captured by nuclei of the U-238 isotope, thus converting it into a fissionable element. Hence breeder reactors produce (or "breed") more fuel than they consume.

The several breeder reactors now in operation throughout the world are primarily prototypal plants with power ratings of 250–500 MWe. Extensive practical use of breeder reactors (operating at about 40% efficiency) is not expected before the late 1980s (33).

PROBLEMS WITH NUCLEAR ENERGY

Most problems concern long-term storage of nuclear waste, possible nuclear power plant accidents that would cause the release of radioactivity, and theft and sabotage of nuclear materials (186). Other issues, including environmental pollution, are not peculiar to nuclear reactors.

A major source of pollution from nuclear plants is waste heat (thermal pollution). A 3 300-MWt reactor produces 3.3×10^9 watts of thermal energy. One thousand million watts are converted into electricity (1 000 MWe); the rest (2 300 MWt) is waste heat (117). The amount of thermal pollution depends on plant efficiency; however, as the efficiency ratings of nuclear and fossil fuel plants are comparable, this is not a controversial issue.

The reactor fuel in a nuclear power plant becomes highly radioactive. If the plant is to operate efficiently, a portion of the reactor fuel must periodically be removed and replaced. Often reactors are designed so that part of the core can be replaced annually. Generally, the fuel element produces power for three or four years before it is shipped to a reprocessing plant (57). Reprocessing removes from the spent fuel those materials which can be re-used in a nuclear reactor.

The remaining materials are "high-level" wastes which require long-term storage and perpetual surveillance. These radioactive wastes are potentially hazardous and expensive to handle. They result only from the chemical reprocessing of used power-plant fuel. The reprocessing of a metric ton of slightly enriched nuclear fuel, which can provide about 200×10^6 kilowatt-hours of electricity, will yield less than half a cubic metre of highly radioactive solid wastes (40).

These wastes contain several materials. Plutonium is highly radioactive and has a 24 000-year half-life. Other long-lived radioactive waste products are strontium and cesium, each with a half-life of about 30 years. These should normally decay to innocuous levels in about a thousand years, whereas plutonium requires several hundred thousand years (88).

There is some question as to whether current technology is capable of providing safe long-term storage and whether future generations should be expected to maintain and safeguard the storage sites.

Another major controversy is the safety of nuclear power plants. Studies have been conducted to determine the possibility of various types of accidents occurring in the reactor. A nuclear reactor cannot explode like an atomic bomb because the content of fissionable U-235 in the fuel is too low. A reactor may get too hot, however, if for some reason the cooling fluids do not remove the heat as fast as it is being generated. If this occurs, there is little risk as long as the fuel is solid; but once it melts there is the possibility of the products of fission escaping into the environment (194).

A 1972 study by the U.S. Atomic Energy Commission estimates that the chance of a melt-down followed by major radioactive leaks is about one in 100 000 per reactor-year. The chance of a really catastrophic accident was set at only one in 10^9 per reactor-year (12), However, the reliability of the report has been questioned.

The predicted number of casualties in the event of the worst possible accident varies considerably. Some believe the casualties of the worst possible melt-down would reach about 3 300, although such variables as wind and population density in the vicinity could alter this figure (197); others maintain that 23 000 to 36 000 persons might die after radiation exposure from a melt-down (61).

Furthermore, special nuclear material could be stolen and diverted for use in nuclear weapons, although the likelihood is an open question. The possibility of sabotage is equally unpredictable.

Nuclear energy production based on fission remains highly controversial. Some question whether it should continue at all; others see it as the only hope for the modern world; and still others regard it as a temporary means until alternative methods are developed.

FUSION

One alternative undergoing considerable research is nuclear fusion. In this process the heating of atoms (e.g., hydrogen) to extremely high temperatures causes them to collide at such velocity that almost all their electrons are stripped off their nuclei. This substance is called a plasma. Bare nuclei can approach one another more closely than whole atoms. At extremely high temperatures they smash into one another and cling together, forming more complex nuclei (11). In thermonuclear reactions (so called because of the intense heat required) a large amount of energy is released: a gram of hydrogen undergoing fusion produces fifteen times as much energy as a gram of uranium undergoing fission (11).

The main fuel for fusion reactions is an isotope of hydrogen known as deuterium. There are an estimated 32×10^{12} metric tons of deuterium in the oceans (11). This supply is considered more than ample and can be obtained and processed rather cheaply.

Fusion reactors are strictly experimental, but cost and efficiency estimates are encouraging. Efficiency ratings of approximately 50% or more (101) — higher than nuclear fission or fossil-fuel plants — would cut down thermal pollution.

The most serious hazard of a fusion reactor is probably fire; however, because of the very small amount of nuclear elements in the reaction chamber at any time, the energy released in a fire would be incidental. There is no danger of thermonuclear explosion. If a fire consumed all components of the system, the total energy released would be about the same as from a large oil tank (101).

The kind and amount of radioactive elements created in a fusion reactor are minor compared with those of a fission reactor. Even so, care must be taken to prevent their release to the atmosphere. It will be difficult to divert these radioactive elements for use in weapons because they are generated, circulated and burned within the fusion plant. Besides, a nuclear weapon cannot be constructed without fissionable material to initiate the explosion (193).

Fuel will be cheap in relation to the energy output. As all the technological problems are not yet known, the cost to the consumer of electricity generated by nuclear fusion cannot be projected. Nevertheless, present estimates indicate that electric power from nuclear fusion will be competitive in price with power from other major sources, such as coal and nuclear fission (101).

Although fusion reactions are elementary processes, many scientific and technological problems must be overcome before controlled energy release on a practical scale can be attained (193). Extremely high temperatures (millions of degrees) must be maintained long enough (0.1 to 1 second) to ignite the fuel atoms and maintain energetic self-sustained fusion reactions (33).

Fusion has been achieved in laboratories throughout the world, but the physical conditions described above have yet to be realized. If it proves to be possible to produce energy from nuclear fusion on a practical scale, the first commercial plants may be built by the late 1990s (101).

Direct conversion of energy

Most power used today is generated by machines with moving (rotating or reciprocating) parts. Falling water turns wheels, steam drives turbine blades, the force of combustion moves pistons — all of this motion is harnessed to produce power by dynamic conversion.

On the contrary, direct conversion transforms energy into useful energy without the aid of moving parts. Direct converters have inherent advantages over dynamic converters: they tend to be simple in design and need little maintenance. Their lack of moving parts lessens the chance of mechanical failure and thus ensures reliability, durability and greater efficiency. Many types of direct conversion devices appear to be ideal for small-scale electrical production in remote areas where they would have to operate reliably for long periods with little or no maintenance; however, most of these generators still present technical and economical problems which limit their practicality.

MAGNETOHYDRODYNAMICS

The principle of magnetohydrodynamic (MHD) generators is relatively simple (Fig. 87). A gas (plasma) consisting of pulverized coal and air heated to over 1 100°C is injected into a chamber where the addition of tiny amounts of chemicals frees the electrons in the gas, which thus becomes electrically charged. The gas then moves at supersonic speeds into another chamber with electrodes on its walls. A strong magnetic field outside the chamber forces the electrons to the electrodes, thereby producing an electric current (65).

A 50% efficiency is anticipated for MHD generators, but efficiency may reach 60% in large plants (52). MHD generators therefore yield

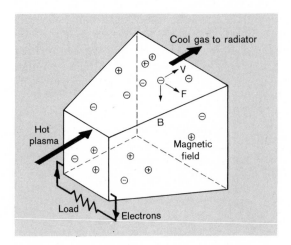

FIGURE 87. Magnetohydrodynamic (MHD) generator. An electrically charged gas is injected into the chamber, where a magnetic field (*B*) forces the electrons to move in direction *F*. The electrons then collect on the right-hand side of the duct and are carried to the load to produce electricity. The remaining gas continues in direction *V*, where the increased volume of the duct reduces the temperature of the gas (47).

up to 50% more power than conventional turbogenerators with the same amount of fuel and with less heat waste (52, 65).

Whereas sulphur emissions are often a problem in coal-burning plants, sulphur is easily precipitated out by the high temperatures in MHD generators. This also permits the use of coal with a high sulphur content in MHD plants (52).

Because MHD generators are still experimental, many problems regarding life-span, temperature and pollution have yet to be solved. Coal ash or alkali residues tend to build up on the flow tube (52). As the magnetic field must be very strong, there is the question of whether the energy produced by the generators is sufficiently more than that required to power the magnets (196).

Japan, the Federal Republic of Germany, the U.S.S.R. and the U.S.A. have experimented with MHD generators (52). In the U.S.S.R., a 25-megawatt generator has actually been tied into a working power grid (65), and a 1 500-megawatt plant is being planned (52). The U.S.A. put a 32-megawatt MHD power system in operation, but it has been dismantled (196). MHD researchers believe that 1 000-megawatt power plants that will be economically and environmentally competitive with conventional power systems can be built (196).

FUEL CELLS

Fuel cells are an interesting combination of technologies. They are similar to chemical batteries, except that conventional fuels are supplied continuously to the cell (25, 216). A number of fuels can be used — natural gas, petroleum products, synthetic liquid or gaseous mixtures made from coal or oil shale (270).

The fuel first goes through a conditioner for conversion into a gaseous mixture of hydrogen and carbon dioxide. This gas and air are then fed into the fuel cell, in which there are electrodes and an electrolyte solution (Fig. 88). The hydrogen from the gas, the oxygen from the air and the electrolyte react so that the anode collects electrons, which are subsequently sent through a load to produce useful work (216).

As fuel cells require no heat input, they are not bound by the typical heat-engine efficiency limitations. Theoretically, an efficiency of 100% would be possible with a pure fuel (216). Typical fuel cells deliver up to 30% more energy per unit of fuel than conventional generating systems (270).

Other advantages include: no emissions, little noise, no coolant, and compactness. A 26-megawatt plant, which could meet the needs of a community of 20 000 people, would be only 5.5 metres high and occupy less than a quarter of a hectare of land (270). Fuel cells, being modular, can be constructed quickly and additional cells can be added at any time.

Fuel cells also lend themselves to small- and medium-scale applications. They could be placed where the demand for power is greatest to reduce transmission costs and losses (270). Currently fuel cells cost $400 per kilowatt capacity and have a life expectancy of 16 000 hours. A manufacturer in the U.S.A. expects to reach the break-even point soon (52).

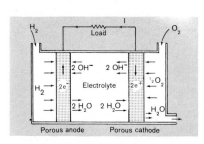

FIGURE 88. Fuel cell diagram. A gas containing hydrogen and a gas containing oxygen are continually injected into the fuel cell, in which there are two electrodes and an electrolyte. The following chemical reactions take place:

At the anode: H_2 (gas) $+ 2 OH^- \rightarrow 2 H_2O + 2e^-$

At the cathode: $\frac{1}{2} O_2$ (gas) $+ H_2O + 2e^- \rightarrow 2 OH^-$

Sum: H_2 (gas) $+ \frac{1}{2} O_2$ (gas) $\rightarrow H_2O$ (liquid)

The electrons leave the fuel molecules at the anode and do work en route to being captured by the reaction at the cathode (216).

Photovoltaic Cells

Photovoltaic cells are powered by solar radiation rather than by fossil fuels. For a detailed description of photovoltaics, see pages 114-117.

Thermoelectricity

Thermoelectricity is closer in principle to photovoltaics than to fuel 'cells or magnetohydrodynamics as it converts thermal energy into electric energy. A thermoelectric converter consists of rods of two different materials (semiconductors) which are joined together at both ends. Each rod is treated with impure substances so that one rod has an excess of electrons and the other a deficiency.

The deficiency creates "positive holes" which travel through the material like electrons, and when one junction of the semiconductors is heated, the electrons and the "positive holes" travel to the colder ends (Fig. 89). The resulting positive and negative terminals form a power source called a thermocouple. Thermoelectric converters supply low-voltage high current, but they can be connected in series, or a thermopile, to obtain higher voltages (47).

The U.S.A. has used thermoelectric converters in space satellites, but

FIGURE 89
Thermoelectric generation. Each semiconductor is treated with impurities to create either an excess or a deficiency of electrons. When the lower junction is heated, the electrons and the "positive holes" (caused by a deficiency of electrons) move to the opposite, colder ends of the semiconductors. The resulting positive and negative terminals provide a power source (47).

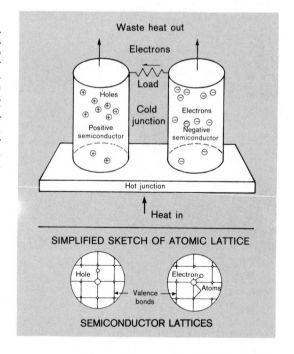

Waste heat out

Electrons

Load

Holes

Cold junction

Electrons

Positive semiconductor

Negative semiconductor

Hot junction

Heat in

SIMPLIFIED SKETCH OF ATOMIC LATTICE

Hole

Electron

Atoms

Valence bonds

SEMICONDUCTOR LATTICES

their present use is minimal. Small generators fuelled by propane are commercially available to operate television sets, and the U.S.S.R. has produced kerosene lamps which use thermoelectric elements (216).

There are several technical obstacles to the further development of thermoelectricity. Each semiconductor is a delicate chemical balance of compounds which must be protected, and the fastening of the rods to the hot junction is difficult. As a result, thermoelectric converters are fragile and costly. Furthermore, the efficiency of thermoelectric conversion is only 1% to 5% (47).

Thermoelectricity does, however, have the important advantages of simplicity of design and potential high reliability because the converter lacks moving parts (47). These features lend themselves to small-scale generation (only a few watts) in remote areas — e.g., for weather stations, buoys and hospital equipment (216). However, owing to the limitations mentioned above, large thermoelectric generators are unlikely to be developed (216).

THERMIONICS

Thermionics is another method for converting heat energy into electrical energy. Two metal plates of different materials — both good conductors — are placed close together. One plate is heated to such a high temperature ($1\,600°$–$2\,000°K$) that its electrons are forced off the surface into a gap between the plates (216). The electrons that collect in the gap eventually repel subsequent electrons back on to the surface from which they came.

To circumvent the repelling forces, the gap is filled with a gas containing positively charged particles, which neutralize the electrons and change the gas into a good conductor. The electrons forced off the heated surface travel through the gas and are absorbed by the other metal plate (Fig. 90).

The key to producing power from thermionic converters is to use a metal for the heated plate that requires more work to force the electrons off the surface than the other plate requires to capture them, hence forming a voltage drop that creates energy.

Theoretically, the efficiency of thermionic generators is as high as 75%; however, at present their efficiency is modest (between 10% and 20%) because of heat losses. On the other hand, they are very reliable, have a long life and can use a variety of heat sources, such as nuclear reactors, decaying isotopes, concentrated sunlight and conventional burners (216).

At present, thermionic generators are used only as small sources of electric power from the heat generated by decaying radioactive

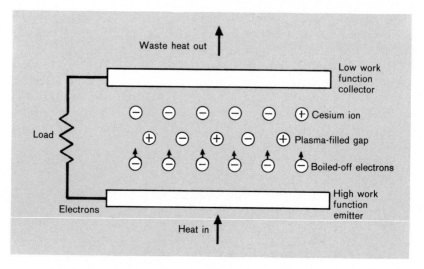

Waste heat out

Low work function collector

⊖ ⊖ ⊖ ⊖ ⊖ ⊕ Cesium ion

⊕ ⊖ ⊕ ⊖ ⊕ Plasma-filled gap

⊖ ⊖ ⊖ ⊖ ⊖ ⊖ Boiled-off electrons

Load

High work function emitter

Electrons

Heat in

Figure 90. Thermionic conversion. The bottom metallic plate is heated to such a high temperature that its electrons are forced off the surface into the gap between the two plates, where there is gas with good conductivity. The electrons flow through the gas and are absorbed by the unheated metal. A voltage drop is formed because more energy is required to force the electrons off the heated surface than is required by the other plate to capture them (47).

isotopes. Large-scale applications will be possible only if better materials are discovered or created (216).

Organic residues

Organic residues consist mainly of carbohydrates: cellulose and lignin, from photosynthesis, or sugar- and food-processing wastes. The conversion of cellulose or lignin into energy provides about 4 500 kilocalories per kilogram of dry matter (oil yields 10 000 kilocalories per kilogram, which is the standard reference for the efficiency of the different processes to be examined).

Another component of these residues is water. The moisture content in residues ranges from a few percent in straw to more than 90% in slurry, sewage or other liquid wastes of the food-processing industry. It is impossible to convert all these materials into energy by the same process.

Another general characteristic of inorganic materials is low density. For instance, one kilocalorie of straw takes up twenty times more space than one kilocalorie of oil. The handling of organic materials for conversion into energy is costly.

AVAILABILITY

There are two major ways to obtain biomass for use as an energy source: (*a*) to collect by-products of wastes and (*b*) to grow a crop for that specific purpose.

In the U.S.A., for example, climate, economics of agricultural production and marketing, and land and water availability all combine to limit the feasibility of growing organic matter for fuel (216). Production costs are low, however, where land is available for nonharvested plants (e.g., the African savanna).

Collecting residue is more practical. Wastes are sometimes concentrated in one place, like industrial by-products (sawdust, bark, food-processing residues, etc.), or are collected for environmental reasons (urban refuse, manure, etc.). In these cases the cost of collection for energy conversion is low or even negative; but often the dispersion and the periodic availability of these raw materials limit their use.

Wood

Wood was probably the first fuel used by man. At the beginning of this century it accounted for about 30% of world energy consumption; today, it still represents 6% (Table 76). The total amount of all woody material used annually for fuel is about 1 400 million cubic metres, the equivalent of 300 million metric tons of oil (136). This is about half of the total world production of wood. Wood provides up to 80% of the energy used in some of the developing countries (181).

Wood is either used directly as fuel after natural drying or transformed into charcoal where shortages exist (202). An estimated 95% of the households in the developing countries where wood is readily available use it as a primary fuel for cooking and heating (31).

Wood by-products

Bark and branches become by-products when wood is processed. Processing also produces large quantities of sawdust. Waste materials supply much of the fuel required by mills, but often they are only incinerated. In the paper industry the burning of solid wood has practically ended, but wood derivatives in the form of a black liquor supply a notable share of the industry's fuel needs — for example, in Finland, 55% (54).

Food-processing refuse

The major food-processing wastes from commodities grown mainly in the hot developing countries are listed in Table 77. In western

TABLE 76. – COMPARISON OF WOOD FUEL AND COMMERCIAL ENERGY
CONSUMPTION THROUGHOUT THE WORLD, 1971

Region	Commercial energy	Wood fuel	Total	Wood fuel as percent of total
	.... 10^{16} joules			
Southeast Asia and Oceania	151	248	399	62
Southern Asia	311	237	548	54
China and the rest of Asia	1 330	132	1 462	9
Near East and north Africa	235	59	294	20
Western and central Africa	32	97	129	75
Eastern and southern Africa	35	105	140	75
Central America and the Caribbean	297	30	327	9
South America	435	178	613	29
Total	2 826	1 086	3 912	28
Developed countries	16 390	143	16 500	1
WORLD TOTAL	19 216	1 229	20 412	6

SOURCE: (258)

Europe, starch wastes are the most important food by-product. As concentrations of wastes must be destroyed anyway, they might just as well be converted into energy.

Crop residues

Table 78 and Figure 91 show that cereal residues (especially straw from rice, wheat, barley and maize) provide good possibilities for energy production, with a potential of 495 million metric tons oil equivalent. Their conversion to energy is currently limited by alternative uses, as for fertilizers or animal feeds, and by collection and transport costs.

TABLE 77. – WORLD ESTIMATES OF SOME FOOD-PROCESSING WASTES, 1975 [1]

By-products	Estimated production (10^6 metric tons)	Average moisture content (percent)	Lower caloric value (kcal/kg)	Oil equivalent (10^6 metric tons)	Present level of energy production
Groundnut shells	5.5	3–10	4 000–4 500	2.3	High [2]
Coffee husks	2.15	13	3 700–3 900	0.8	Low
Bagasse (cane)	110.0	40–50	2 000–2 500	24.0	High
Cotton husks	6.3	5–10	4 000	2.5	High
Coconut husks	13.0	5–10	4 000	4.2	Low [3]
Cocoa cabosse	9.5	80–85	—	—	None
Rice hulls	55.0	9–11	3 300–3 600	18.0	Low [4]
Olives	3.0	15–18	4 000	1.2	Average
Oil-palm ears	3.4	70	—	—	None
Oil-palm fibres	2.9	55	1 800–2 000	0.5	High
Oil-palm husks	2.0	6	4 000–4 500	0.8	High

[1] Based on world production of the given commodities and current by-product ratios from various sources collected in CEEMAT. – [2] Half of the estimated production in oil-processing plants. – [3] In small factories, as most coconuts are directly consumed. – [4] High in rice mills, but the major part of rice production is directly consumed.

Nonharvested plants

In equatorial forests or tropical savanna (where elephant grass grows wild), more than 15 tons of dry matter per hectare are produced annually through photosynthesis.

Animal excreta

Most animal species excrete daily about 1 to 1.5 kilograms of dry matter per 100 kilograms of live body weight. Most excreta, often mixed with agricultural by-products to form compost, are returned to the land to improve the physical and chemical structure of the soil.

However, in many developing countries with serious wood shortages, cow dung and other non-dried excreta are burned as fuel. In India,

TABLE 78. – ESTIMATED WORLD PRODUCTION OF CEREAL STRAWS WITH OIL EQUIVALENTS, 1974 [1]

Cereal crops	Straw residues	Oil equivalent of dry matter
	.... 10^6 *metric tons*	
Wheat	360	126
Paddy rice	323	113
Barley	171	60
Maize	586	102
Rye	33	11
Oats	51	18
Millet	92	32
Sorghum	94	33
Total	1 710	495

[1] Adapted from *FAO production yearbook* assuming a straw-grain ratio of 1:1 in dry matter and a moisture content of about 15–20% for all straws but maize (55–60%).

FIGURE 91
Estimated production of straws by world regions, 1974.

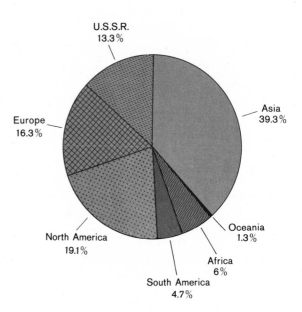

U.S.S.R.
13.3%

Asia
39.3%

Europe
16.3%

North America
19.1%

Oceania
1.3%

Africa
6%

South America
4.7%

for example, dung and agricultural wastes account for about 22% and 34%, respectively, of the total energy consumption (136). Through biogas, it is possible to extract energy from manure without much loss of fertilizer value.

Urban refuse and sewage

The collection of urban refuse and sewage is necessary for environmental and sanitary reasons. The quantities are roughly proportional to the standard of living. In the U.S.A. the annual production of dry matter is estimated at about 130 million tons in refuse and 12 million in sewage — the energy equivalent of 26 million metric tons of oil (8).

Urban refuse has a moisture content of about 50–60%, and gives off 1 500–2 000 kilocalories per kilogram when burned. (The calorific value per kilogram is considerably higher in the developed countries.) In some large cities, urban refuse is used to produce energy. An alternative use, composting, may be more attractive for cities with less than 100 000 inhabitants (103), whereas incineration with energy recovery is probably more attractive for larger cities. Owing to the high dilution of the dry content, energy can be recovered from sewage only in a second stage — the mud treatment — by fermentation-generating methane.

Ocean kelp

The harvesting of plants from the sea has enormous potential. The giant kelp, a seaweed, is abundant, accessible and capable of biological conversion to methane. It commonly reaches 70 metres in length and grows two thirds of a metre or more per day (178). It is native to waters with average temperatures of less than 20°C.

Kelp bed densities vary in different parts of the world. Biomasses with a fresh weight of up to 22 kilograms per square metre have been reported in California beds, while Indian Ocean biomasses average 140 kilograms per square metre (124, 280).

Fresh kelp is about 85% water and consists of about 70% volatile matter on a yearly average. A simple digestion system yields 0.41 cubic metre of methane per kilogram of digestible organic matter.

Of the 36×10^7 square kilometres of ocean on the earth, only 56–70% is arable surface water. A kelp farm system 1 120 kilometres in diameter could supply the equivalent of the total energy consumption of the U.S.A.

Dry processes

Energy conversion can lead directly to heat production or to intermediate fuels (solid, liquid or gas). The two major methods of conversion are (1) dry processes, in which the material is transformed under high temperatures, and (2) wet processes, which involve biological processes such as fermentation.

The reaction needs air, or oxygen, and gives off heat. Like all chemical reactions, suitable concentrations of reagents (fuel, oxygen) are needed. The speed of the reaction is highly influenced by temperature, and the reactions may be endothermic (needing heat) or exothermic (giving off heat).

When heated, an organic material begins to lose its water content; the higher the initial moisture content, the more heat is needed. The dissociation of the products begins at about 250°C and becomes somewhat more rapid at higher temperatures (e.g., 500°C). Typically, this dissociation yields the following proportions of the initial heating value: a solid residue called char, consisting mainly of carbon, 42%; tar or oil mixed with water, 30%; and gases having a heat value of about 3 000–4 000 kilocalories per cubic metre (350–450 Btus per cubic foot). The proportions depend, however, on the processing conditions.

The heat required for dissociation, as well as for initial removal of the moisture, may be produced by combustion *in situ* of some of the gases if a little air is added. Otherwise heat is supplied by pyrolysis, the process used to obtain coke from coal and charcoal from wood.

The gases, generally burned to recover heat, could be used for power generation or as a starting material for chemical production (ammonia, methanol). The oil/tar phase contains (1) a heavy part which can be treated to recover its value (oils), giving a residue that is usually recycled for a thermal cracking, and (2) an aqueous part which can be distilled to recover such products as aromatics, methanol and various oils. Char is recovered and sold as a fuel with a caloric value of about 6 000–8 000 kilocalories per kilogram depending on its ash content. The process does not necessarily end at this stage. Char, for instance, is very active and may be gasified if it reacts with steam:

$$C + H_2O + 32 \text{ kilocalories} \longrightarrow CO + H_2$$

This reaction starts at a high temperature (800°C) and requires considerable heat because it is endothermic.

Tar may be dissociated in the same way. There is always enough steam because of the initial moisture content and the first volatile gases. The reaction is induced by moving the steam and providing heat. Heat may be produced by partial combustion of the gases in the presence of oxygen or air. Other reactions can also occur (Fig. 92), such as the following:

$$C + CO_2 + 41 \text{ kilocalories} \longrightarrow 2CO$$

At this point no more solids or liquids remain, except mineral ashes, and because of the high temperature, only gases are given off. These gases may be burned directly if oxygen continues to be supplied. Their heating value depends on the conditions of the process, as the more air that has been introduced to give heat by partial combustion of the gases, the higher the proportion of noncombustible gases (N_2

FIGURE 92. Dry-process conversions.

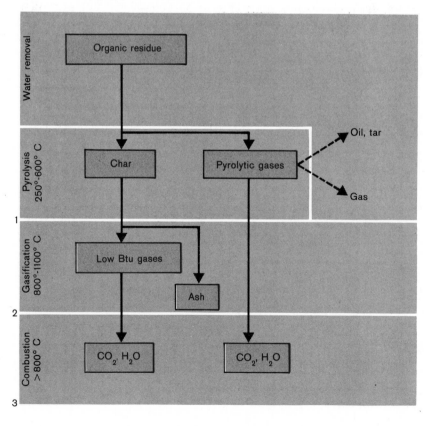

coming from the air, CO_2 and H_2O). An indicative value is 1 000–1 200 kilocalories per cubic metre (1 050 Btus per cubic foot).

Combustion

Combustion is the most common form of energy conversion, mainly because of the widespread burning of wood and animal excreta. It is a series of many chemical reactions, each following the other as the appropriate conditions make them possible. Usually, heat produced by combustion is used for heating, domestic purposes or industrial applications. The hot water or steam released during industrial processes can be used to produce electricity and for heat.

In the simplest burners, wood or any kind of organic material is put on grates or on the ground, and the natural convection of air reduces it. A great deal of heat is wasted, however, because combustion is incomplete — that is, some carbon remains in the ashes.

In large-scale applications, the material is burned on grates with a controlled introduction of air. In well-designed burners, primary air (10–50% of the air, depending on the initial moisture content) is introduced under the grates to make the material gaseous; then secondary air is added above the bed (or in a special nozzle) to burn the gases, thereby permitting high-temperature, clean combustion. In poor designs, all of the air is introduced under the grates, and the presence of smoke indicates low efficiency.

The area of the grates is calculated to avoid the removal of volatile ashes. Depending on the product, usually 200–400 kilograms of organic residue can be burned per hour on a grate that is one metre square. Steam pipes are placed above the ignition bed for heat exchange. Otherwise, hot gases are introduced in a boiler.

Smaller, more sophisticated designs — for example, cyclonic and fluidized bed types — have been developed for the production of powdered products by suspension burning (279). Another possibility is to co-fire the organic residue with a solid fuel, such as coal.

Pyrolysis

Commercial pyrolysis processes were developed mainly for municipal waste, in which the moisture content ranges from 30% to 70%. A few plants are in operation in the U.S.A., and the feasibility of pyrolytic treatment of agricultural waste has been studied.

These rather complicated processes have long been used in charcoal production, mainly from wood (202). With the simple, traditional methods of making charcoal, all the pyrolytic gases are lost except those burned *in situ* to generate the heat needed; thus, only 25–40% of the initial calorific value is recovered in charcoal. In more sophisti-

cated retorts, distillation is conducted to recover volatile products
in the gases.

Most charcoal is produced in kilns. The better types of kiln require
6 to 8 cubic metres of wood per ton of charcoal; earthen kilns require
up to 12 cubic metres per ton. Charcoal is used primarily for cooking
and heating. Despite the energy loss during the conversion of wood
into charcoal, it costs less than fuelwood to transport beyond about
30 kilometres (136).

Gas producers

Char and tar are converted at high temperatures into gaseous products. Because a lot of heat is needed for gasification of most of the
organic material, some air is introduced to cause partial combustion
in situ.

The gases produced, called low Btu gases because their calorific
value is 1 000–1 200 kilocalories per cubic metre, consist mainly of
N_2 (from the air introduced), CO and H_2 (Table 79). Because of
their low calorific value, it is not economical to transport them long
distances; therefore, it is best to use direct combustion processes.

Gas producers are most commonly used in conjunction with engines
to generate mechanical power. For technological reasons (high temperature, dust) the gases must be cooled and cleaned before being
introduced into the engine, which may be a spark-ignition engine with
adapted compression ratios or a Diesel (dual-fuel) engine. In the

TABLE 79. — TYPICAL GAS ANALYSIS FROM
A DOWNDRAFT GAS PRODUCER USING WOOD

Gas	Percent by volume
Carbon dioxide (CO_2)	9.5–9.7
Hydrocarbons	0–0.3
Oxygen (O_2)	0.6–1.4
Carbon monoxide (CO)	20.5–22.2
Hydrogen (H_2)	12.3–15.0
Methane (CH_4)	2.4–3.4
Nitrogen (N_2)	50.0–53.8

SOURCE: (118)

latter, low Btu gases and fuel oil contribute, respectively, about 90%
and 10% of the power.

Gas producers for both stationary and mobile engines were first
introduced in the 1940s owing to the fuel shortages caused by the war.
Perhaps 700 000 vehicles were adapted to gas generators using various
fuels (wood, charcoal, coal, coke, coconut husks). They were similar
in design to that shown in Figure 93. Filtration, a major problem
on mobile engines, was easily solved on stationary generators by the
use of scrubbers.

Although gas producers have not been improved much since then,
an interesting application called the Delacotte process was recently
developed in France (Fig. 94). The two stages of the process are
separated: pyrolysis takes place in the upper part of the chamber,
gasification in the lower part. The pyrolytic gases burn in an auxiliary
chamber.

The plant may be constructed in two ways: for charcoal and gas
production with minimum air feeding or for gas generation with all
possible intermediates. The Jura plant has two dual-fuel engines
generating electricity at 600 kilovolt-amperes each. Normally about
0.8 kilogram of dry wood and 25 grams of fuel oil are consumed per
kilowatt-hour produced (177).

FIGURE 93
Automotive wood-gas gen-
erator, ca. 1945. Equiv-
alent: 25 pounds of
wood — 1 gallon of gas-
olene; 2.5 kilograms of
wood — 1 litre of petrol
(161).

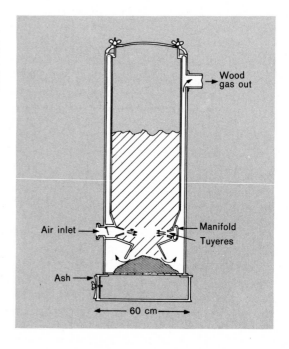

Wood
gas out

Air inlet

Manifold

Tuyeres

Ash

60 cm

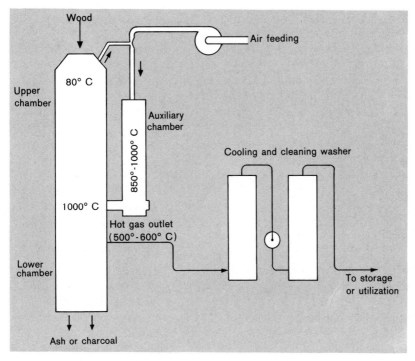

FIGURE 94. Diagram of the Delacotte system of producing wood gas (62).

The conversion of dry organic residue to cold, clean gas is 65–75% efficient (118). The generation of electricity with an alternator and a dual-fuel engine is about 35% efficient, delivering a combined efficiency of nearly 25%, compared with the 30–35% efficiency obtained when electricity is produced with a boiler and steam.

WET PROCESSES

Another major method of converting organic materials into energy is by biological means, such as fermentation, with eventual chemical treatment. Because microorganisms are involved, wet processes differ from dry processes in two ways: (1) an aqueous environment is necessary, and (2) they are slow and depend on the nutrient conditions.

Through fermentation, organic residues can be converted into liquid or gaseous fuels. The two most common means are anaerobic fermentation and alcoholic fermentation.

Anaerobic fermentation to produce methane

The direct digestion of organic residues by bacteria in an aqueous environment provides a combustible gas: methane (175, 200, 205, 209,

225, 244). Furthermore, the residues from this process can be returned to the land as a soil conditioner. For this reason, anaerobic fermentation has provoked considerable interest in many countries, although it is used mainly in India and the Republic of Korea.

Most organic materials provide methane by anaerobic fermentation under the proper conditions. Anaerobic digestion consists of three phases (208): (1) enzymatic hydrolysis; (2) organic acid formation (with bacteria); and (3) methane generation (*if* the materials are methane formers). Thus the methane process requires a symbiosis of acid-forming and methane-forming bacteria. The reactions of the two groups must occur simultaneously; if they become unbalanced, the digestion process fails. This explanation schematizes the real process — not well known — which by many paths and biochemical reactions leads to the formation from complex organic wastes of the end product, methane.

Because bacteria are living organisms, methane production is influenced by two main factors: (1) nutrient components of the substrate composition and (2) temperature.

Many different organic waste materials have been used as food material for anaerobic digestion. Almost any organic waste can serve as a component of the substrate, but the bacteria can use as food only what is digestible for them. As carbon and nitrogen are the main nutrients of the bacteria, the suitability of materials for gas production is based on the carbon to nitrogen (C/N) ratio, which should be about 30 for optimum production.

Recent improvements in anaerobic digestion have been made with the use of various animal manures as a substrate. The use of crop residues is less practical because their C/N ratio must be reduced by excrements for satisfactory biogas production.

Temperature is the other determining factor. There are two sets of methane-forming bacteria: (1) the mesophilic organisms, for which the optimum temperature is 35°C, and (2) the thermophilic organisms, for which the optimum temperature is 55°C. Technologically the mesophyllic process is the most simple. Methane production usually decreases rather drastically as the temperature drops below optimum level. A widely used rule of thumb is that methane production drops 50% for every 11°C decrease in temperature.

Poisoning agents and pH values must also be considered. Operating skill is required for fermentation control. Generally, adequate fermentation rates can be achieved with a pH of 6.7 to 7.6 (154). A frequent and inconvenient problem arises when the digester becomes too acid during the start-up period owing to an excess acid production over methane production.

Digester outputs. Theoretically, the fermentation of cellulose should give off equal amounts of methane and carbon dioxide; but, actually, part of the CO_2 produced remains fixed on an organic or a mineral basis or it is dissolved in the water. Table 80 lists the composition ranges expected in biogas.

The biogas produced in a well-functioning digester may have a heat value of 5 500 kilocalories per cubic metre (600 Btus per cubic foot). By comparison, natural gas has a heat value of 9 500 kilocalories per cubic metre (1 000 Btus per cubic foot). The total gas mixture, including impurities, may be used to produce heat, as in stoves or space heaters, or for work in internal-combustion engines. If, however, steps are taken to lessen the CO_2 content, the energy value of the gas per unit volume increases. The CO_2 content can be reduced by bubbling the gas through a water scrubber. To decrease corrosiveness, the hydrogen sulphide (H_2S) can be removed by a filter (45). Mobile engines or, more commonly, stationary engines for pumping irrigation water or for electric-power generation can use methane, but because of the required compression ratio, they must be designed for this application.

The amount of gas produced is highly variable, depending on temperature, composition of the organic material, dilution and digester design (146). An output of 0.3 cubic metre of gas per kilogram of dry matter is considered excellent (185, 220), although better outputs are reported from specially grown plants, like water hyacinth (52). For practical reasons, gas yields are usually not so high (132, 195, 208). A more common value appears to be 0.15–0.2 cubic metre per kilogram of dry matter — an efficiency of 20–25%. The main reasons

TABLE 80. – GENERAL COMPOSITION OF
BIOGAS PRODUCED FROM FARM WASTES

Gas	Percent
Methane (CH_4)	55–65
Carbon dioxide (CO_2)	35–45
Nitrogen (N_2)	0–3
Hydrogen (H_2)	0–1
Oxygen (O_2)	0–1
Hydrogen sulphide (H_2S)	Trace

for this low yield are that (1) there is always an indigestible part and (2) the long process is stopped before the end of fermentation.

Sludge, or fermented organic material, is the other digester output. Its necessary removal is costly, but it does have fertilizer value as it retains most nutrients, including nitrogen, phosphorus and potash. It can also be used to improve soil structure, an important consideration in many tropical areas. Where manure or by-products are not usually returned to the soil, the introduction of biogas techniques can initiate agronomic improvements.

Equipment. Methane production requires an oxygen-free digester (i.e., a tightly sealed tank), input-output equipment, and a gas collector. Sometimes, a controlled heat input system, agitation, scum prevention and gas treatment are also needed.

The batch process is simpler than the continuous-flow digester, but the gas production is not uniform. Starting from zero when the material is first introduced, the output slowly rises to a peak and then tapers off. Although the batch process is used mostly for animal wastes, it is also suitable for the fermentation of fibrous materials in stacks. In this case, the materials are placed in gas-proof tanks, which are covered during the anaerobic fermentation phase. Several stacks may be affiliated for more continuous gas production.

Continuous-flow digesters include input-output equipment, and the system is set up for periodic (often daily) removal of a portion of the slurry (Fig. 95). Agitation and scum prevention are generally necessary. Gas production is far more uniform if reliable input-output equipment is provided. Most digesters are designed for a retention period of twenty to thirty days. The wastes are diluted with water to facilitate input.

All kinds of materials can be used in the construction of digesters. In the villages of southern and eastern Asia, where many are in use, constructions of local materials are preferred to commercial designs. There, digesters generally consist of a pit lined with stones and concrete or clay and a roofed cylindrical gas holder. The amount of insulation depends on the design. In areas with cold winters a heating system is essential. A family using about 1.5 cubic metres of gas daily will require a digester with a capacity of 3 cubic metres.

The main disadvantages of continuous-flow digesters are high water consumption (a major inconvenience in dry areas), cleaning problems (deposits, scum and crusts) and the skill required for control. Furthermore, they are not suitably designed for fibrous materials or for cold areas (208).

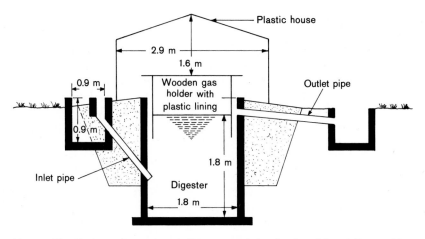

FIGURE 95. Design of a type of digester used in the Republic of Korea (251).

Applications and feasibility. Most digesters are found in the developing countries and use animal excreta, especially cow dung. Because of the shortage of wood fuel in those countries, sun-dried dung is largely used as fuel for direct combustion, which means that all its fertilizer value is lost. Agronomic improvements and increased food production depend on the use of organic residues as fertilizer because tropical soils are often fragile and commercial fertilizers lacking. Therefore, the introduction of biogas is one solution to the energy and fertilizer problem.

Digesters are, however, still too costly for most rural communities. In India, a family-size digester (4–6 cubic metres) costs about $600, an investment beyond individual family means without government help. The present policy encourages village-scale digesters, which are cheaper per unit of production. The cost of energy on this scale is comparable to electricity at six cents per kilowatt-hour (151). Makhijani (151) points out that considering the foreign-exchange requirements for other energy sources, energy from locally produced methane is cheaper. Furthermore, the use of digesters can provide a sanitary means of human waste disposal and, at the same time, conserve valuable nutrients for use on cropland.

The situation is different in the developed countries. In Japan, small digesters were in use after the war, but they were abandoned as commercial fuels and fertilizers became available. There, manures are carefully returned to the soil.

Many authorities doubt that methane digesters are economically feasible for individual farms in western countries; however, numerous firms are successfully marketing anaerobic digesters in the United Kingdom, where treatment of effluent before discharge or payment of a premium to the local water authority is required by law. Increasing fuel prices combined with strict pollution controls may economically justify the use of digesters on certain intensive animal farms. In China, approximately 250 000 small digesters have been reported and the installation of more units has been programmed.

Studies in the U.S.A. conclude that the installation of digesters on farms is uneconomical at the present level of technology and energy costs (74, 221). However, these studies have charged the entire cost of a digester and its operation against energy production, whereas if animal waste is used as the substrate, a portion of the cost might be charged to pollution control and waste handling.

Studies in the developed countries show that many dry processes are better adapted to agricultural by-products than biological means because of higher efficiencies and lower processing costs (35).

Municipal sewage treatment merits special attention. After aerobic treatment, decanted muds can be submitted to anaerobic fermentation, which in the most modern plants produces enough methane to power a dual-fuel Diesel engine that will furnish part or all of the electricity needed to operate the entire plant — and in many cases even a surplus.

Alcoholic fermentation

The sugars present in many agricultural plants can be transformed into alcohol by fermentation — a common process in the food and beverage industry. The starch content can also be transformed into ethyl alcohol, with an intermediate enzymatic action that generates sugars. The "vegetal alcohols" produced throughout the world are thus derived from products rich in sugars or starch, such as sugarcane, sugar beet, grape, molasses, cassava, potato and maize.

Organic residues do not generally have a high sugar or starch content. The molecules of lignin and cellulose must be converted into sugars by hydrolysis. Two methods are used: (1) acid hydrolysis, with or without heating (glucose yields are about half the weight of the cellulose), and (2) enzymatic conversion of cellulose derived from a fungus, which is reported to give a better glucose yield as all the cellulose in the organic material is transformed (224). Thereafter, the process continues through fermentation of the sugars and alcohol distillation until only an aqueous residue composed mainly of lignin remains.

Ethyl alcohol is a good liquid fuel with a caloric value of about 5 600 kilocalories per litre. It can be used in spark-ignition engines either alone or mixed with gasolene without much change in the power output (114). Only minor adaptations are required on a commercial vehicle.

A recent study on acid hydrolysis of straw reports an efficiency of 24% in the best trials (17, 141). Therefore, technical improvements in hydrolysis would be worthwhile. At present, only one large plant in the U.S.S.R. uses this process.

Alcohol production from sugar- or starch-rich plants is more widespread. In some countries it is a means of absorbing surplus crops. Nevertheless, vegetal alcohol is not yet economically competitive with alcohol obtained from petroleum (17, 161).

Brazil has developed a large-scale programme of alcohol production by cropping sugarcane and cassava for this purpose (191). Typical yields are 66 litres of alcohol per ton of sugarcane (or about 3 000 litres per hectare). Cassava (which can be planted on poor land) provides 180 litres per ton (or about 2 200 litres per hectare). In Brazil about 200 000 cubic metres of vegetal alcohol are already being used annually as engine fuel by mixing it with gasolene (80% gasolene, 20% alcohol). National plans aim for the production of 2 300 000 cubic metres of alcohol, mainly from sugarcane.

Conclusion

The type of energy conversion process chosen for obtaining fuel or heat from organic residues will depend on (1) energy needs, (2) where the organic residues are available and (3) size of the conversion plant. The generally high collection and transport costs for these materials may favour transformation into intermediate fuels or into electricity if a distribution network exists.

A major concern in choosing a process is conversion efficiency (Table 81). For example, as the initial moisture content goes up, an increasing part of the energy in dry processes is used to remove the water. Available dry processes are therefore not appropriate if the initial moisture content exceeds 50–60% because only about half of the potential energy is recovered.

For wet products (70–90% moisture content) only biological means can be used, although sludge may present an environmental problem. Wet processes require more volume for a given production because the process is slow and the material is aqueous. Suitable nutrients and skill are needed to ensure the microorganism's metabolism requirements. Another possibility is to reduce the initial water content in the products so that the more efficient dry processes can be used. This can be accomplished by sun-drying, mechanical dewatering (filter

TABLE 81. – COMPARATIVE CONVERSION EFFICIENCIES OF VARIOUS
PROCESSES APPLIED TO A DRY ORGANIC BY-PRODUCT [1]

Process	Final state of energy		
	Intermediate fuels	Heat	Mechanical power or electricity
	.... *Percent efficiency*		
Combustion	—	65–95	10
Pyrolysis	30–90	—	—
Gas producer	65–75	—	20–25
Anaerobic fermentation	20–35	20–35	5–12
Alcoholic fermentation .	20	20	5

[1] Straw with 15% moisture content.

press, screw press), or artificial drying with heat that would otherwise
be lost (e.g., engine gases).

Research and improvements are still needed for most processes, especially enzymatic hydrolysis, pyrolysis and the design of gas producers.

Geothermal energy

Near the centre of the earth there is a molten core consisting mainly
of iron and nickel. Temperatures inside the core are believed to reach
3 500°–4500°C (64). Most of the heat is produced by decaying radioactive materials within the core; a negligible amount is produced by
friction resulting from solar and lunar tides, the motion of crustal
plates and the compression of subsurface material (216). The heat
spreads out from the core through the mantle and into the crust, with
a temperature gradient ranging from 8° to 15°C per kilometre (216).

The heat in the top 16 kilometres of the earth's crust totals about
13×10^{26} joules — 2 000 times the amount of heat that would be
produced if the world's entire reserve of coal were burned (115). If
only one tenth of the geothermal energy in about the top 3 kilometres
of the earth's crust were extracted and converted into electricity with
today's techniques, it could provide 58 000 megawatts annually for at
least fifty years (115).

FIGURE 96. Major geothermal areas of the world (67).

Countless problems arise in the use of geothermal energy. At present, drilling is limited to a depth of about 12 000 metres (216). Most of the heat within that range is extremely diffuse and cannot be extracted economically, although research on this problem continues. In France, for example, geothermal heat forced from within the earth's crust to the surface is being used to heat about 3 000 lodgings near Paris (222).

Near the surface, however, there are concentrated heat zones associated with the boundaries of the major crustal plates and other geologic features (Fig. 96). These geothermal systems were formed in various ways. Some consist mainly of steam, some contain hot water and still others are composed of dry, hot rock.

FORMATION

Liquid-dominated systems

These geothermal zones are found in volcanic and mountain-building areas. Basically, they are formed by rain water which circulates down into the earth and is heated at a depth of a few kilometres by hot rocks, causing it to expand and move upward. During this process the aqueous solution picks up impurities, including sodium, potassium, sulphate, silica and borate (16).

Vapour-dominated systems

These systems were developed thousands to tens of thousands of years ago by the circulation of water down to an area of intense heat. Eventually, however, little additional water could enter the area to replace that which had turned into vapour; hence, vapour-dominated reservoirs developed. The steam contains other gases (CO_2, H_2S, NH_3) but little or no water (16). Vapour-dominated systems are about twenty times less common than liquid-dominated systems.

Dry, hot rock systems

In dry, hot rock systems, as in vapour- and liquid-dominated systems, the heat produced inside the earth convects upward. Concentrated heat zones can occur near the surface where there are geologic faults in the earth. Dry, hot rock systems form in areas where impermeable rock abounds and underground water is absent. Much energy has been stored in these areas through the ages. The thermal energy released by cooling one cubic kilometre of a dry, hot rock system from 350°C to 177°C has been estimated as equivalent to the energy available in 300 million barrels of oil (16). Unfortunately, a technology for economical cooling has not yet been developed.

TECHNOLOGY AND USE

Although dry, hot rock geothermal systems cannot yet be exploited, geothermal energy from both liquid and vapour systems has been commercially developed in some areas. Reservoirs containing mostly vapour are used to generate electricity, while those in which liquid predominates are often used for their heating value.

Generation of electricity

Electricity was first generated from geothermal steam in 1904 at Larderello, Italy. A 250-kilowatt station established there in 1913 now has a capacity of about 406 megawatts (67). Table 82 lists the world's geothermal power installations currently in operation or at an advanced stage of development. The average temperature of the steam ranges from 170°C to 300°C, and most systems contain a mixture of steam and water. Average drilling depths are 500–1 500 metres.

The geysers in the U.S.A., Italy and Japan are dry-steam geothermal wells (no water particles are present). The technology for using geo-thermal dry steam is similar to that of modern steam-electric plants except for pressure and temperature ranges. Steam from natural steam wells reaches the turbine at a pressure of about 7 bars and a

TABLE 82. – GEOTHERMAL POWER INSTALLATIONS IN OPERATION OR AT AN ADVANCED STAGE OF DEVELOPMENT

Country and site	Average (maximum) drillhole depth (m)	Average (maximum) temp. (°C)	Discharge type (S=steam; W=water)	Total dissolved solids in water (g/kg)	Total generating capacity (MW)	
					Installed	Planned addition
Chile:						
El Tatio	650 (900)	230 (260)	S + W	15	—	15
El Salvador:						
Ahuachapan	1 000 (1 400)	230 (250)	S + W	20	30	50
Iceland:						
Namafjall	1 000 (1 400)	250 (280)	S + W	1	2.5	—
Italy:						
Larderello	600 (1 600)	200 (260)	S	—	406	—
Mount Amiata	750 (1 500)	170 (190)	S (+ W)	—	25	—
Japan:						
Matsukawa, N. Honshu	1 000 (1 500)	220 (270)	S	—	20	—
Otake, Kyushu	500 (1 500)	230 (250)	S + W	2.5	11	—
N. Hachimantai ...	800 (1 700)	— (>200)	S + W	—	10	—
Hatchobaru, Kyushu	1 000 —	250 (300)	S + W	5.5	—	· 50
Onikobe, Honshu .	300 (1 350)	— (288)	S (shallow) + W (deep)	1.5	—	20
Mexico:						
Cerro Prieto	800 (2 600)	300 (370)	S + W	17	75	75
New Zealand:						
Wairakei	800 (2 300)	230 (260)	S + W	4.5	192	—
Kawerau	800 (1 100)	250 (285)	S + W	3.5	10	—

(*continues*)

TABLE 82. — GEOTHERMAL POWER INSTALLATIONS IN OPERATION OR AT
AN ADVANCED STAGE OF DEVELOPMENT *(concluded)*

Country and site	Average (maximum) drillhole depth (m)	Average (maximum) temp. (°C)	Discharge type (S=steam; W=water)	Total dissolved solids in water (g/kg)	Total generating capacity (MW)	
					Installed	Planned addition
Broadlands	1 100 (2 420)	255 (300)	S + W	4	—	100
Philippines: Tiwi, S. Luzon	920 (2 300)	—	S + W	—	—	20.5
Turkey: Kizildere	700 (1 000)	190 (220)	S + W	5	—	10
U.S.A.: The Geysers	1 500 (2 900)	250 (285)	S	—	600	300
U.S.S.R.: Pauzhetsk	— (800)	185 (200)	S + W	3	5	7

SOURCE: (67)

temperature of about 200°C, whereas the pressure and temperature of steam used in a modern fossil-fuel power station are over 200 bars and 540°C. Consequently, the efficiency of geothermal plants is only 14% as compared with 40% for fossil-fuel plants (64).

To generate electricity from wet-steam geothermal wells (containing both steam and liquid), the liquid must be separated out before the steam can be used to operate a turbine in the same manner as that from dry-steam geothermal wells.

About 15% of the electric power in New Zealand is generated from hot-water geothermal sources (216). Steam is obtained by drilling down about 610 metres to a reservoir. The release of pressure through the drill holes causes the hot water to boil and vaporize (215); even so, about 80% (by weight) of the discharge from the drill holes remains unvaporized. The hot water is separated from the steam and discharged into a lower pressure area, where some of it flashes into steam. By reflashing the hot water two or three times, more steam is produced, but pressure is reduced to about 0.07 kilogram per square centimetre which means it can only be used in low-pressure turbines (215). The remaining water is discharged into streams.

Geothermal energy may be the cleanest energy source available, but many factors have yet to be studied. Environmental problems associated with geothermal power plants are high noise levels and gaseous emissions. Moreover, changes in fluid pressures may alter normal fault activity.

Direct use of heated water

In Iceland, water from liquid-dominated geothermal systems is used directly for space heating. In Reykjavik, Iceland, nine out of ten homes are heated by geothermal water, and it is expected that by 1980 about 60% of the nation's population will be using municipal geothermal water supplies for heating (216). The deepest geothermal well in Iceland is about 2 000 metres, and temperatures in the reservoirs range from about 100°C to 150°C (64). The hot water is transported in conduits and pipes for distances up to 16 kilometres and is delivered to homes at about 80°C.

The U.S.S.R. has developed municipal geothermal heating installations which heat homes and furnish hot water for communities of 15 000 to 18 000 people (64). The geothermal energy used for direct heating in the U.S.S.R. may be equivalent to over one million metric tons of fuel oil per year (64).

In Hungary, geothermal hot water is used extensively for heating greenhouses and farm buildings and for drying crops, as well as for space heating in homes and commercial buildings. In several other countries, including Japan, New Zealand and the U.S.A., geothermal hot water has proved economical for space heating, sulphur recovery, desalination, therapeutic spas, etc.

One disadvantage is that the use of geothermally heated water is limited to the immediate vicinity of the wells. Another problem arises from the minerals in the water, as residues can clog equipment and some are highly corrosive. Regardless of the technical difficulties, it appears that geothermal resources will continue to provide a small part of the world's energy supply.

Human energy expenditure

In many areas, especially the tropics, food production relies almost entirely on human muscle power, often with the aid of very simple tools. The estimated animal energy efficiency of humans is about 2.5%, compared with 3–5% for bullocks (151). Man's work efficiency is less than 10% at relatively high temperatures and humidities (233).

The reason for this low efficiency is that the human system is unable to dissipate heat through evaporation, radiation or convection under those conditions.

The laws of thermodynamics constrain humans and animals to low levels of conversion efficiency from food to mechanical work in a tropical climate. Therefore, auxiliary energy sources may have a greater impact on food production in the tropics than under temperate conditions (219).

There are various methods of determining the rate of human energy expenditure. Nearly a century ago, a calorimeter was used. Later, a portable respirometer was developed to measure oxygen consumption — an indirect indicator of energy expenditure.

Dietary studies based on the known calorie content of foods can establish the balance between energy consumption and expenditure. Energy output can also be calculated from the metabolic cost of engaging in various activities for a recorded period of time; or the heartbeat rate can be measured and correlated with the rate of energy expenditure (184, 233, 234).

Studies show that most humans can continuously sustain a power output of only 75 watts and that only 20–30% of the chemical energy in food can be converted into mechanical energy in the form of physical work (49, 230). The other 70–80% goes into maintenance of body growth, recovery from illness and waste heat.

Many measurements have been made of the chemical energy expenditure of humans walking on a level surface. Figure 97 shows results from five investigations (184). In the range of 3 to 6.5 kilometres per hour, energy expenditure is linearly proportional to speed. At greater speeds, the requirement increases at higher rates. Of course, the rate of energy expenditure is also proportional to body weight. Figure 98 illustrates the rate of chemical energy expenditure while walking on a treadmill at various gradients.

Many agricultural tasks involve walking on soft or uneven surfaces. Table 83 shows the effect of different surfaces on the rate of chemical energy expenditure. Note that 35% more energy is required to walk on a ploughed field than on an asphalt road (184).

Other agricultural tasks involve carrying loads. Figure 99 shows the results of a study in which a person weighing 56 kilograms walked 100 metres, picked up a load and returned with it at a speed of 4.5 kilometres per hour. The loads were carried eight different ways. With a yoke across the shoulders the rate of energy expenditure was least. When the load was carried on the hip and under the arm, it was higher. Intermediate values were obtained when the load was carried in trays, in hand bundles, on the head and over the shoulder (184).

• Atzler, E. & Herbst, R. *Pflügers Arch. ges. Physiol.,* 215: 290-328, 1927.

⸰ Benedict, F.G. & Murschhauser, H. *Energy transformations during horizontal walking.* Carnegie Institution of Washington , 1915

▪ Brezina, E. & Kolmer, W. *Biochem. Z.,* 38: 129-153, 1912.

⸰ Douglas, C.G. & Haldane, J.S. *J. Physiol.,* 45: 235-238, 1912

▫ Margaria, R. *Mems Accad. naz. Lincei,* 7: 299-368, 1938.

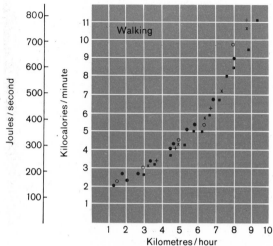

FIGURE 97. Rate of human chemical energy expenditure for walking on a level surface at different speeds (184).

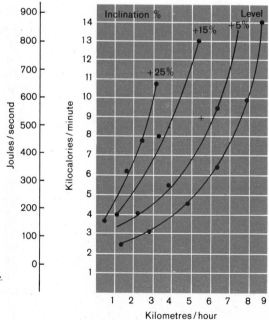

⸰ Keys, A. *et al. Biology of human starvation.* University of Minnesota Press, 1950.

● Margaria, R. *Mems Accad. naz. Lincei,* 7: 299-368, 1938

FIGURE 98. Rate of human chemical energy expenditure for walking up 5%, 15% and 25% grades and on a level surface (184).

TABLE 83. – RATE OF HUMAN CHEMICAL ENERGY EXPENDITURE FOR
WALKING ON VARIOUS SURFACES

Surface	Speed	Energy cost	
	Km/h	Kcal/min	Joules/s
Asphalt road	5.5	5.6	391
Grass track	5.6	6.2	433
Potato furrows	5.4	6.9	481
Stubble field	5.2	6.8	474
Ploughed field	5.3	7.6	530

SOURCE: (184)

The effect of speed on the rate of energy expenditure when carrying
loads of different weights is shown in Figure 100. At slow speeds the
rate of energy expenditure is low, but at speeds above 4 kilometres per
hour the energy required increases rapidly.

Hand tillage involves digging or spading. Examples of chemical
energy requirements for pick, spade and shovel work are shown in
Table 84. The usual energy expenditure rate for shovelling is about
6 or 7 kilocalories per minute (450 or 520 watts) — extremely heavy
work that cannot be sustained more than a few minutes without rest.
Tables 85–90 give energy expenditure rates for various other agricultural
tasks and working conditions.

For accurate assessment of manual power capabilities, the rate
of energy expenditure must be observed over extended periods of time.
One-week data for coal mining operations, which should be comparable
in energy requirements to many agricultural tasks, are presented in
Table 91. The total caloric intake for the week was 26 460 kilocalories,
and the energy used while working was 11 076, or 42% — a high conversion rate (184).

Coal mining with little mechanization is extremely hard work.
This is demonstrated in Table 91 by the 7 hours and 57 minutes per
week of using food energy at the rate of 6.7 kilocalories per minute
(about 520 watts) for hewing and walking, along with other high-
energy tasks that only the strongest, most physically fit humans can

FIGURE 99
Rate of human chemical energy expenditure for carrying loads in various ways. Although not shown in the graph, the energy requirements for carrying a load in trays, on the head and over the shoulder were intermediate between those for carrying a load with a yoke and on the hip (184).

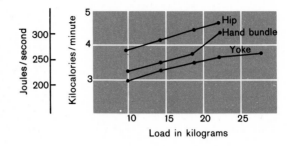

TABLE 84. – RATE OF HUMAN CHEMICAL EXPENDITURE OF MEN WORKING WITH PICK, SPADE, SHOVEL AND WHEELBARROW

Activity	Energy cost	
	Kcal/min	Joules/s
Shovelling 8-kg load distance of 1 m:		
Less than 1-m lift, 12 throws/min	7.5	523
From 1- to 2-m lift, 12 throws/min	9.5	663
Shovelling 8-kg load distance of 2 m:		
Less than 1-m lift, 10 throws/min	8.5	593
From 1- to 2-m lift, 10 throws/min	10.5	733
Digging trenches in clay soil	8.5	593
Shovelling 8-kg load distance of 1 m:		
0.5-m lift, sand, 12 throws/min	5.4	377
0.5-m lift, gravel, 12 throws/min	7.2	502
1.0-m lift, sand, 12 throws/min	6.0	419
1.0-m lift, gravel, 12 throws/min	8.4	586
1.5-m lift, sand, 10 throws/min	6.0	419
1.5-m lift, gravel, 10 throws/min	8.0	558
Shovelling	6.0	419
Digging	6.8	474
Hewing with pick	7.0	488
Pushing wheelbarrow with 100-kg load	5.0	349
Pushing wheelbarrow at 4.5 km/h on fairly smooth surface:		
57-kg load	5.0	349
150-kg load	7.0	488

SOURCE: (184)

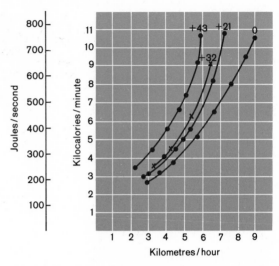

FIGURE 100. Rate of human chemical energy expenditure for carrying loads at different speeds. The curves with dots are for loads carried high on the back in a knapsack supported by straps. The curve with crosses represents soldiers carrying military equipment (184).

TABLE 85. – RATE OF HUMAN CHEMICAL ENERGY EXPENDITURE IN PLOUGHING, WEEDING, BINDING AND THRESHING, U.S.S.R., 1933

Sex	Age	Weight (kg)	Activity	Mean and range	
				Kcal/min	Joules/s
M	24	65	Ploughing	6.9 (6.3–7.8)	481 (440–544)
M	21	67	Ploughing	5.4 (5.2–5.6)	377 (363–391)
M	24	65	Threshing rye	5.0 (4.8–5.2)	349 (335–363)
M	32	76	Threshing rye	4.5 (4.2–4.9)	314 (293–342)
F	19	63	Binding oats	3.3 (3.0–3.8)	230 (209–265)
F	19	52	Binding oats	4.1 (3.6–4.7)	286 (251–328)
F	19	63	Binding rye	4.2 (3.8–4.9)	293 (265–342)
F	19	52	Binding rye	4.7 (4.1–5.1)	328 (286–356)
F	19	52	Weeding rape	3.3 (3.0–3.5)	230 (209–244)

SOURCE: (184)

TABLE 86. – RATE OF HUMAN CHEMICAL ENERGY EXPENDITURE IN MOWING, SHOCKING AND BINDING GRAIN, HUNGARY, 1932

Number of subjects	Number of experiments	Activity	Mean and range	
			Kcal/min	Joules/s
6	20	Mowing wheat	7.7 (6.2–10.2)	537 (433–712)
2	9	Mowing barley	7.0 (5.6– 8.5)	488 (391–593)
4	12	Preparation of shocks	6.6 (5.1– 8.4)	460 (356–586)
3	12	Binding wheat	7.3 (6.2– 8.6)	509 (433–600)

SOURCE: (184)

TABLE 87. – RATE OF HUMAN CHEMICAL ENERGY EXPENDITURE IN MOWING, BINDING, LOADING AND THRESHING GRAIN, ITALY, 1945

Number and sex of subjects with age range	Activity	Kcal/min (mean)	Joules/s (mean)
15 men (15–55)	Mowing with a scythe	6.8	474
15 men (15–25)	Mowing with a horse-drawn reaper	4.3	300
8 women (13–29)	Preparing bundles of cut maize	5.0	349
11 men (15–57)	Binding bundles of maize	6.8	474
8 men (19–41)	Preparation of shocks	5.5	384
4 women (14–29)	Preparation of shocks	4.8	335
5 men (16–41)	Loading shocks on carts	5.6	391
7 men (15–41)	Threshing: Throwing sheaves from pile to ground	6.0	419
7 men (15–41)	Throwing sheaves toward threshing machine	5.1	356
7 men (15–41)	Throwing sheaves up to threshing machine	5.8	405
5 women (13–25)	Binding sheaves	3.8	265
5 women (13–25)	Throwing sheaves into machine	5.5	384

SOURCE: (184)

TABLE 88. – RATE OF HUMAN CHEMICAL ENERGY EXPENDITURE IN LAND CLEARING, RIDGING, PLANTING, WEEDING AND POUNDING RICE, THE GAMBIA, 1953

Number of subjects	Sex	Age	Mean weight and range (kg)	Effective temperature (°C)	Number of observations	Activity	Mean and range	
							Kcal/min	Joules/s
4	M	25–38	63 (56–68)	17–21	11	Clearing shrub and dry grass	7.1 (5.8–8.4)	495 (405–586)
5	M	24–36	57 (49–68)	21–25	13	Ridging (deep digging)	9.5 (5.5–15.2)	663 (384–1 060)
4	M	21–28	67 (64–70)	23–27	12	Planting groundnuts	3.7 (3.1–4.5)	258 (216–314)
3	M	19–33	63 (61–66)	17–22	9	Weeding	5.3 (3.8–7.8)	370 (265–544)
3	F	18–30	54 (48–58)	23–27	11	Hoeing	5.8 (4.8–6.8)	405 (335–474)
4	F	18–28	57 (43–65)	24–29	15	Pounding rice	5.0 (3.9–6.4)	349 (272–447)

SOURCE: (184)

TABLE 89. – RATE OF HUMAN CHEMICAL ENERGY EXPENDITURE IN AGRICULTURAL OPERATIONS, NIGERIA, 1954

Six subjects, mean wt. 55 kg

Number of observations	Activity	Mean energy cost	
		Kcal/min	Joules/s
12	Grass cutting	4.3	300
12	Bush clearing	6.1	426
12	Hoeing	4.4	307
12	Head carrying, 20-kg load	3.5	244
12	Log carrying	3.4	237
10	Tree felling	8.2	572

SOURCE: (184)

TABLE 90. – RATE OF HUMAN CHEMICAL ENERGY EXPENDITURE IN PLOUGHING AND MILKING OPERATIONS, GERMANY (F.R.), 1953

Subject	Activity	Mean and range	
		Kcal/min	Joules/s
Age 28, wt. 64 kg	Milking by hand	4.7 (3.5–6.3)	328 (244–440)
	Machine milking one pail	3.4 (3.2–3.7)	237 (223–258)
	Machine milking two pails	3.9 (3.1–4.6)	272 (216–321)
	Cleaning milk pails	4.4 (4.3–4.5)	307 (300–314)
Ages 18–39, wt. 57–86 kg			
16 observations	Horse ploughing	5.9	412
6 observations	Horse ploughing (another type of plough)	5.1	356
8 observations	Tractor ploughing	4.2	293
14 observations	Tractor ploughing (another type of tractor)	4.2	293

SOURCE: (184)

TABLE 91. – RATE OF HUMAN CHEMICAL ENERGY EXPENDITURE OF A COAL MINER OVER ONE WEEK, GREAT BRITAIN, 1952

Male, age 32, ht. 1.77 m, wt. 67 kg. Occupation: stripper

Activity	Total time		Energy cost		Total	
	H	Min	Kcal/ min	Joules/ s	Kcal	10⁴ joules
In bed	58	30	0.94	66	3 690	1 545
Recreational and off-work:						
Light sedentary activities .	38	37	1.59	111	3 680	1 540
Washing, shaving, dressing	5	3	3.3	230	1 000	419
Walking	15		4.9	342	4 410	1 846
Standing	2	16	1.8	126	250	105
Cycling	2	25	6.6	460	960	402
Gardening	2		5.0	349	600	251
Total recreational and off-work	65	21			10 900	4 563
Working:						
Loading	12	6	6.3	440	4 570	1 913
Hewing	1	14	6.7	467	500	209
Timbering	6	51	5.7	398	2 340	980
Walking	6	43	6.7	467	2 700	1 130
Standing	2	6	1.3	91	230	96
Sitting	15	9	1.63	114	1 530	640
Total working	44	9			11 870	4 968
TOTAL	168				26 460	11 076
Daily average	24				3 780	1 582
Food intake (daily average determined from diet survey)					3 990	1 670

SOURCE: (184)

accomplish. Presumably the work periods were often interrupted by short breaks.

A person can perform considerable work over time. With a diet of 3 000 kilocalories per day, at a 20% conversion rate, 600 kilocalories or nearly one horsepower-hour can be transformed into mechanical work. If the hourly rate of exertion is one-tenth horsepower, one horsepower-hour of work can be accomplished in a ten-hour day.

A human adult totally at rest for twenty-four hours still requires over 1 000 kilocalories, depending on body weight, just for maintenance. Any external activity not only requires additional food energy but also increases the maintenance requirement (102).

Pesticides

Pesticides — herbicides, insecticides and fungicides — are used to protect agricultural crops. They are essentially substitutes for the human and mechanical labour otherwise needed to control weeds, insects and fungus.

ENERGY CONSIDERATIONS

The manufacture of chemical pesticides requires considerable energy. The energy inputs for several herbicides have been calculated in the United Kingdom. Table 92 compares the results with the energy inputs for several fertilizers. (The data are believed to be within 15% accurate.) Although the total energy content of herbicides is about ten times that of fertilizers, the typical application rates per hectare on cereals are 250 kilograms of fertilizer and only 0.75 kilogram of her-

TABLE 92. – COMPARISON OF ENERGY INPUTS FOR THE MANUFACTURE
OF HERBICIDES AND FERTILIZERS [1]
(10^9 joules/metric ton)

Herbicides [2]	Naph-tha	Fuel oil	Natu-ral gas	Elec-tricity	Steam	Total	Fertilizers [3]	Total
MCPA	53.3	12.6	12.0	27.5	22.3	130	Nitram	25
Diuron	92.3	5.2	63.1	85.6	28.3	270	Urea	35
Atrazine	43.2	14.4	68.8	37.2	24.7	190	NPK	
Trifluralin	56.4	7.9	12.8	57.7	16.1	150	(17–17–17)	18
Paraquat	76.1	4.0	68.4	141.6	169.3	460		

SOURCE: (105)

[1] For reference, the inherent energy of coal is 25 × 10^9 joules/metric ton. – [2] 100% active ingredient. – [3] Including packing and bagging.

bicide (105). Therefore, the total energy input for one application of a herbicide (200 × 10⁹ joules per metric ton) is about one thirtieth that for one application of a fertilizer (20 × 10⁹ joules per metric ton).

Although the manufacture of chemical pesticides is energy intensive, their effect on yields and other crop production energy requirements must be taken into account. The use of herbicides reduces the need for hand and mechanical tillage and cultivation, although they can be, and in many areas are, alternatives (5). The benefits also lower costs and energy requirements for fertilizers, irrigation, harvesting, drying, transport and storage. Moreover, crop yield losses and consequently the amount of land needed for crop production are reduced. Also, fewer labourers are needed.

Energy inputs for various weed control methods are shown in Table 93. Hand labour uses less energy than most mechanical methods and is comparable to some forms of herbicidal weed control; however, with the exception of hand labour, more than one type of weed control is generally used. Table 94 is an energy balance sheet for various weed control methods in maize. Hand labour has the best output-input energy ratio, but the damage to small grains planted as seed caused by the hoe and the hoer can be extensive. Herbicides have the crucial advantage of quickly halting weed infestation.

DEMAND

Insecticides are still the most widely used pesticides in developing countries, but in regions with substantial fertilizer use and high-yielding plant varieties, herbicides are becoming increasingly important. In fact, herbicide use in the developing countries has increased more rapidly than expected. The reason is apparent from data for the developing countries showing 30–80% lower wheat yields without herbicide treatment (78); the results for rainfed rice in Africa and Asia are similar. Worldwide herbicide use has continued to increase rapidly, as seen in Table 95, and they now represent over half of all pesticide production (78). Consequently, herbicides are in short supply.

SUPPLY

Insecticides are used for agriculture and for public health purposes. Because about half of all insecticides are used on cotton crops (78), the supply-demand situation is closely linked to the textile industry. Much cropland in the U.S.A. is planted with cotton, but many farmers could raise alternative crops if textile prices drop. If this should happen,

TABLE 93. – ENERGY INPUTS FOR VARIOUS WEED CONTROL PRACTICES PER PERFORMANCE

Method \ Input	Gasolene	Indirect machine	Hand labour	Herbicide	Total
	.... Kilocalories/hectare				
Hand labour			53 800		53 800
Field cultivator	120 800	60 400	170		181 370
Tandem disk	93 100	46 500	220		139 820
Rod weeder	26 000	13 000	170		39 170
Rotary hoe	19 700	9 800	120		29 620
Row cultivator	36 700	18 300	310		55 310
Rotary tiller	262 300	131 100	930		394 330
Herbicide, 0.5 kg/ha	8 100	4 000	70	13 600	25 770
Herbicide, 1 kg/ha	8 100	4 000	70	27 200	39 370
Herbicide, 2 kg/ha	8 100	4 000	70	54 400	66 570
Herbicide, 4.5 kg/ha	8 100	4 000	70	108 700	120 870

SOURCE: (167)

TABLE 94. – ENERGY BALANCE SHEET FOR VARIOUS WEED CONTROL METHODS IN MAIZE

Input \ Method	Cultivation	Cultivation plus herbicide	Herbicide	Cultivation plus hand labour	Hand labour
	.... Kilocalories/hectare				
Chemical	—	27 000	82 000	—	—
Mechanical	137 000	142 000	12 000	137 000	—
Hand labour	1 000	1 000	—	24 000	81 000
TOTAL	138 000	170 000	94 000	161 000	81 000
Hours of hand labour/hectare	1.4	1.5	0.1	43.5	89.0
Output/input ratio	49:1	54:1	96:1	57:1	117:1

SOURCE: (167)

TABLE 95.– RATES OF GROWTH IN DEMAND FOR PESTICIDES WITH DISTRIBUTION BETWEEN DOMESTIC SOURCES AND IMPORTS FOR DEVELOPING COUNTRIES, 1971–73 AND 1975–77

	Number of countries	1971–73				1975–77			
		1971 base	Compound rate of growth per annum	Source of increased supplies		1975 base	Compound rate of growth per annum	Source of increased supplies	
				Domestic	Imported			Domestic	Imported
		Thousand metric tons		*Percent*		*Thousand metric tons*		*Percent*	
North and Central America	3								
Herbicide		1.5	19	50	50	2.0	10	50	50
Insecticide		17.1	19	42	58	28.1	11	21	79
Fungicide		4.3	3	100	0	4.9	7	87	13
Total		22.9	17	45	55	35.0	10	30	70
South America	7								
Herbicide		3.4	23	18	82	5.7	8	50	50
Insecticide		12.9	17	0	100	33.0	3	12	88
Fungicide		9.8	28	79	21	33.1	2	43	57
Total		26.1	22	41	59	71.8	3	31	69
Asia and Oceania	15								
Herbicide		4.1	35	42	58	6.2	17	13	87
Insecticide		28.7	36	52	48	54.0	12	85	15
Fungicide		10.1	31	40	60	20.3	10	81	19
Total		42.9	35	49	51	80.5	12	76	24
Africa	13								
Herbicide		0.4	108	0	100	2.6	76	0	100
Insecticide		14.4	–11	0	100	11.3	11	0	100
Fungicide		0.6	9	0	100	0.7	41	0	100
Total		15.4	–6	0	100	14.6	27	0	100
All countries	38								
Herbicide		9.4	32	29	71	16.5	25	12	88
Insecticide		73.1	21	47	53	126.4	9	53	47
Fungicide		24.8	25	61	39	59.0	5	63	37
Total		107.3	23	48	52	201.9	9	46	54

SOURCE: (78)

it is unlikely that the demand for insecticides would exceed the supply.

It appears that sufficient insecticide manufacturing capacity has been planned to ease shortages associated with demand increases anticipated in the near future (78). Nevertheless, the rapidly growing demand for fungicides and the shortage of feedstock required in their manufacture have led to an insufficient supply. Long range supply-demand situations are difficult to forecast.

So many variables are involved that only rough predictions can be made for future pesticide supplies. Generally, increased production is planned, but the industry is afflicted with many problems. One major difficulty is the higher manufacturing cost caused by the depletion in fossil fuels. (For example, the price of DDT increased two and a half times from mid-1973 to the beginning of 1975.) Shipping capacity shortages may continue to be a problem in the future as they have been in the past.

In Europe, registration and certification problems have caused some new pesticide manufacturing plants to remain idle for up to a year after completion (78). Perhaps the major problem associated with insecticides is the increasing vector resistance to some chemicals. Research is still needed to develop better, more efficient pesticides and to eliminate hazardous effects of the chemicals on the environment.

The lives, let alone the welfare, of hundreds of millions of people even now depend upon a continuing and adequate supply of the correct pesticides, in the right place, at the right time, at a price that is not inflated by shortages and excessive demands [79].

Hydropower

Small rivers and streams can provide limited supplemental power for driving machinery or generating electricity. Hydropower systems are reliable and flexible, and they require only simple equipment and little maintenance. The system must be adapted to the site. The changes in water flow with the seasons and hydraulic conditions must also be considered when designing the system (43).

The theoretical power provided by flowing water is calculated as

$$P \text{ (kW)} = 9.8 \times Q \text{ (m}^3\text{/s)} \times H \text{ (m)}$$

where Q is the quantity of water flowing past a point per second (estimated by multiplying the cross-sectional area of the stream by the stream velocity), and H is the height from which the water falls.

TABLE 96. – CHARACTERISTICS OF WATERWHEELS AND SOME WATER TURBINES

Type	Flow rates (litres/second)	Range of waterhead (metres)	Revolutions per minute	Percent efficiency	Power output (kilowatts)	Ability to handle changing conditions		Technological level of construction
						Flow	Head	
Waterwheels:								
Undershot	300–3 000	0.5–3	2–12	40–75	1–10	Good	Fair	Low/medium
Overshot	30–1 000	3–10	2–12	50–80	1–10	Good	None	Low
Water turbines (microplants only):								
Pelton	30–500	50–1 000	500–1 000	82–85	40–400	Good	None	Medium/high
Mitchell (Banki)	100–3 000	3–50	100–400	80–84	30–700	Good	Good	Medium
Francis	50–5 000	10–200	250–1 000	82–90	100–1 000	Medium	Poor	High
Kaplan or propeller	1 000–15 000	2–20	200–500	80–90	20–1 000	Medium	Medium	High

WATERWHEELS

Waterwheels date back to Biblical times and are still widely used throughout the world. They are often more economical for small power requirements (1–10 kilowatts) where no other generating facilities are readily available.

Waterwheels are serviceable where fluctuations in flow rate are large but speed regulation is not practical. They are generally used for slow-speed applications, as for flour mills, some agricultural equipment and pumping (43).

Undershot or overshot wheels convert water to mechanical power. The characteristics of waterwheels and of water turbines (from which the generally higher power output is transformed into electricity) are given in Table 96.

WATER TURBINES

Water turbines are needed to achieve a higher power range (20 kilowatts to a megawatt or more). They can be connected to the general electrical distribution network, but they are especially valuable in isolated areas and small towns where a local network exists. The characteristics of the major types of water turbines are outlined in Table 96.

Hydroelectric sites are classified according to their waterhead into three main categories requiring different types of water turbines:

1. High heads (more than 150 metres). The water is directed by a nozzle into bucket-shaped cups mounted on the perimeter of a wheel known as a Pelton turbine. The capital costs, as opposed to maintenance costs, comprise the larger share of the final cost per kilowatt-hour.

2. Average heads (15–150 metres). The water intake constructions are similar to those used with high heads, but the turbines are designed for less water flow. The Francis and Mitchell turbines are included in this category. In the Francis turbine the water is distributed over the entire surface of a paddle wheel, similar to the wheel of a centrifugal pump or fan, and moves from the periphery to the interior, thereby generating power. Sometimes the Francis turbine is placed in the water chamber. In the Mitchell type the water is introduced at one point.

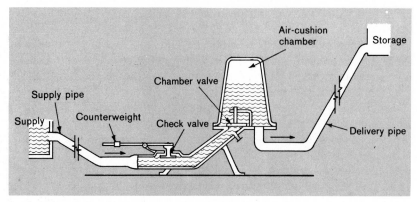

FIGURE 101. Operating principle of the hydraulic ram (271).

3. Low heads (2–15 metres). The propeller (fixed position) and Kaplan (movable) turbines with blades are among those used with low heads. Often the turbines are connected directly to the alternator, and they are placed in a pipe or in a siphon.

Alternating current is usually generated with water turbines, but flow regulation is required. A technical solution can be found for each specific problem, but the cost per kilowatt-hour throughout the year must be carefully scrutinized.

Generally, hydro equipment costs more initially than a Diesel engine performing the same task, but this is offset by lower operating costs for the hydro system (43).

HYDRAULIC RAM

A hydraulic ram uses power from falling water to force a small portion of the water to a greater height than the source; only about one part in ten of the water is delivered (275). The principle of operation is illustrated in Figure 101. Water flows through the supply pipe and the open check valve. As the water velocity increases, the pressure on the underside of the check valve rises until it overcomes the weight of the valve, which causes it to move rapidly upward, thereby closing the opening. The pressure of the moving water opens the chamber valve so that water is driven into the air-cushion chamber, where the increasing air pressure causes it to be discharged through the delivery pipe.

As the momentum of the water in the ram decreases, the chamber valve closes. This creates a sudden drop in pressure under the check valve, causing it to drop down and open quickly, and the cycle begins again. The frequency at which the cycle is repeated is regulated by adjusting the counterweight (4).

A ram works at about 75–90 strokes per minute and is about 60% efficient (4). For example, water falling at a rate of 10 litres per second from a height of 5 metres can be raised by a hydraulic ram to a height of 55 metres at 0.6 litre per second.

If enough water is available and the waterhead is high enough, hydraulic rams are practical pumps. No external power is needed and, except for the initial installation, the cost is negligible.

4. ALTERNATIVE STRATEGIES

Although the developing countries strive to maintain a series of balances — for example, between exports and imports, population and food supply, resource development and utilization — the evidence suggests that they have real problems of integration and that their schisms and internal conflicts prevent them from devising rational strategies with respect to energy. A change in energy policy usually means treading on the interests of an economic or a political interest group; hence, energy questions are settled in the realm of politics rather than in a rational and objective context.

For example, the petroleum embargo of 1973 and the subsequent price increases constitute a discontinuity in the evolution of the economies of nations that import or export petroleum. Because of the higher costs of many vital commodities and services, the "ripple" effect of petroleum price increases is still being felt. Although experts disagree on future petroleum price projections, further price increases are probable because the world supply is finite and the demand continues to grow.

What range of options or alternative strategies should be considered by planners and by agricultural and energy technologists? Can a balance between energy and food supplies be maintained as the world population continues to grow and petroleum supplies remain uncertain? If there is insufficient petroleum to meet all the demands, what sort of priorities should be established? Should price be the primary determinant of allocation or should other mechanisms be considered?

On a global scale, how can poor nations compete with rich ones for increasingly expensive petroleum products? What alternatives are there to petroleum as an energy base? Will alternative energy resources be cost effective? These and many other complex and difficult questions must be answered if all nations are to thrive and world order is to be maintained.

Increasing the supply of energy

To adequately feed the expanding population and meet other social and economic goals, the amount of energy used per person per hectare in agricultural production, processing and distribution should be greatly increased from present levels. The extension of low-energy techniques now commonly used — the man or woman with only a hoe, or with bullock and plough, and ox-cart transport to the nearest village — will not produce the surpluses needed to feed rapidly growing urban populations, often far removed from areas of agricultural production.

Poor countries need as much energy as they can get, as cheaply as possible. Previously, development policies in most developing countries were able to ignore proper management and effective use of "noncommercial" energy sources because cheap oil was available and technology was usually borrowed from industrialized countries which had failed to develop and use "noncommercial" sources of energy (151).

The use of biomass for energy is discussed in Chapter 3. Other sections of Chapter 3 explain conversion processes and principles that might be employed to help increase the supply of both commercial and noncommercial energy, particularly in the developing countries. Detailed cost-benefit analyses are needed to determine the feasibility of each process under particular conditions.

Revelle in his analysis of energy use in rural India (201) points out that a considerable increase in energy use will be necessary to meet future food needs — primarily for irrigation, chemical fertilizers and additional draught power for cultivating fields.

The climate and water supply permit two crops per year on most of India's arable land, but this will be possible only if facilities for surface and groundwater irrigation are greatly expanded and if an abundance of nitrogen fertilizers can be made available so that the fields do not have to be left fallow to accumulate nitrogen. According to estimates of the Indian Irrigation Commission, for full irrigation development about 46 million net hectare-metres should be pumped annually from wells, which would require at least 1×10^{14} kilocalories of fuel energy — four times the bullock, Diesel and electric energy now being used. Applications of nitrogen fertilizer should be raised to about 100 kilograms per hectare per crop, or 20 million metric tons for 100 million double-cropped hectares, with an energy requirement of 3.5×10^{14} kilocalories.

More draught power than can be obtained from bullocks is needed for rapid seedbed preparation to grow two crops a year. Makhijani and Poole (151) estimate that an additional 5×10^5 kilocalories per hectare per crop are required to construct and operate small tractors, or 1×10^{14} kilocalories if 100 million hectares are double-cropped.

The cultivation of two crops per year would greatly increase farm employment, probably by at least 50%, corresponding to an added human energy input of 0.3×10^{14} kilocalories per year. The total additional energy requirement would be 5.8×10^{14} kilocalories more than the energy now used in Indian agriculture. With these added energy inputs it would be possible, in principle, to approximate the average U.S. yield of 3.28 metric tons per hectare per crop for food-grains as compared with the present Indian yield of 0.8 ton.

Assuming a yield of 2 tons per hectare per crop and double-cropping on 100 million hectares, foodgrain production could be raised to between 300 and 400 million metric tons (depending on the amount of farmland devoted to other crops), or between three and four times present production. The input of energy from all sources per ton of foodgrains would be on the order of 1.8×10^6 kilocalories, significantly less than at present. If the average yields per hectare in U.S. foodgrain production were attained, the energy input would be about 1.1×10^6 kilocalories per ton of foodgrains.

Revelle (201) concludes: "The men and women of rural India are tied to poverty and misery because they use too little energy and use it inefficiently, and nearly all they use is secured by their own physical efforts. A transformation of rural Indian society could be brought about by increasing quantity and improving the technology of energy use." This conclusion is applicable to virtually all the non-oil-producing developing countries; however, among the obstacles are landlord-tenant relations, the lack of political pull for allocating national resources to agriculture and the inequality of educational opportunities between city and countryside (131).

Conversely, in the developed countries the generalization that increased energy use per caput leads to a higher standard of living is being questioned. Perhaps the law of diminishing returns is limiting the rate of increase in production. Sweden and Switzerland, for example, have achieved a high standard of living with one half to two thirds the energy consumption rate per caput of the U.S.A. The energy savings in Sweden and Switzerland can be attributed largely to multifamily dwellings and the extensive use of mass-transit systems.

MORE EFFICIENT ENERGY USE

It has been suggested that agricultural units (and machines) are too large in the U.S.A. and other developed countries (42, 44, 139, 157, 211). Certainly waste and inefficiency should be minimized, and techniques for expanding production with smaller increments of energy sought more vigorously. As agriculture, like all other sectors, will ultimately have to survive without petroleum, planning for a gradual shift in the energy base should begin now.

Much research has been devoted to increasing the efficiency of agricultural production (169, 170). For example, better water management, fertilizer use, and machinery design, selection and operation are constantly being sought. Several general categories of efficiency improvement are:

(*a*) reduction of waste without new investment through better management and farming practices;
(*b*) capital investment in new, more efficient technology; and
(*c*) changes in crop-livestock mix and consumption patterns from high- to low-energy systems.

Although priority must be given to increasing food production, some energy inputs can be reduced. Energy flow diagrams (see Chapter 2) illustrate the food production areas where energy ratios could be improved significantly:

(*a*) better photosynthesis through the use of more energy-efficient varieties or better agricultural practices;
(*b*) reduced harvesting and storage losses for grain and forage through good preservation techniques;
(*c*) improved conversion of grain and forage into animal products with selected breeds and good management practices; and
(*d*) emphasis on less sophisticated food, such as cereals and vegetables.

Changing consumer habits and improving food processing are effective ways to conserve energy. The great potential for increased photosynthetic efficiency must be studied by agronomists and biochemists. But these are long-run prospects. In the short run it is necessary to strive for increased efficiency in agricultural production. Energy conservation in agriculture often consists in better management of all necessary operations or elimination of unnecessary activities (34).

The major uses of energy in agricultural production can be classified as indirect (fertilizers and machinery) and direct (heat and power from fuels).

FERTILIZER USE

Increased fertilizer use is primarily responsible for rising food production. Yield responses to fertilizer, however, depend on good soil preparation, adequate water and seed selection. This response to fertilizer inputs is expressed in an asymptotic curve. Fertilizer inputs in many developed countries have almost reached the peak of the curve. If application rates are too high, fertilizer is wasted.

There are two approaches to reducing consumption:

1. Fertilizer application only where and when the plant needs it; spreaders which provide accurate distribution (e.g., centrifugal spreaders); placement near the plant, especially when it is young; for nitrogen, several small applications rather than a single large application, especially on grasslands; fertilization at the right moment, to ensure maximum plant use and to avoid leaching losses.

2. Better utilization of residues as crop by-products or manure and higher nitrogen fixation with selected plants or through inter-cropping; improved manure-handling systems and recycling.

MACHINERY AND ASSOCIATED FUEL NEEDS

Machines that will perform their function more effectively with reduced losses and lower energy inputs must be conceived. Improved designs are especially needed to increase the efficiencies of pumps and motors. Tractor and other stationary and mobile engines are now primarily Diesel-powered. Other less efficient types are gradually being phased out in some countries, as in the U.S.A. Although tractor-engine efficiency remains about the same, the weight per horsepower output has somewhat decreased.

It is important to keep tractor engines properly adjusted for good fuel efficiency, and they should also be used for optimum loads. Implement size should be compatible with the power unit. As tractors are often used for tillage, agronomic methods that reduce soil movement will save energy (22, 68).

Alternative fuels, utilization of waste, heat recovery, new materials, plant and animal genetics, and educational programmes on energy conservation should also be considered.

IRRIGATION

Energy could be saved in irrigated agriculture by reconsidering sprinkler and gated pipe systems, devising new harvest techniques and using soil amendments to improve water retention. The use of in-dustrial and municipal waste water should be studied. Improved management methods for energy conservation would include computer scheduling of irrigation for better satisfaction of crop requirements.

FUELS FOR HEATING

Heating fuels, the third large energy input, are used mainly in crop drying and greenhouses and, to a lesser extent, in livestock buildings and processing. Alternative sources and either waste heat from industrial plants or heat pumps may likewise be used effectively.

Industrial waste heat has been used successfully in greenhouses. The main problems are the necessity of locating the greenhouse near the plant rejecting heat and the low water temperatures. Heat pumps may be used in greenhouses not requiring high temperatures if this equipment can be used throughout most of the year.

In drying operations, better drier design for appropriate air-flow rate and temperature at each drying stage must be considered. For grain drying a technique known as "dryeration," consisting of slow ventilation without heating in the final stage, could be more widely used.

Wet processes, such as silage, for crop preservation may be preferable to drying for livestock feeding on a farm. Another alternative is mechanical dewatering when the initial product (e.g., grass or sugar beet pulp) is very wet.

FOOD PROCESSING

In the U.S.A. where 12–20% of energy consumption is associated with the food system, the potential for energy saving or for replacement of petroleum and natural gas by less critical fuels is significant. As the price of petroleum and natural gas increases relative to other inputs, and as regulations to promote energy conservation evolve, the U.S. food system can be expected to respond by implementing the results of the many energy conservation studies now under way.

Recommended energy conservation projects for production agriculture and the agricultural processing industry are listed in Tables 97 and 98.

Energy conservation and elimination of waste

There is some potential for saving energy in the food systems of the developing countries, but the developed countries have far greater opportunities for conservation. Two workshops sponsored by the U.S. Energy Research and Development Administration considered this question (70, 71).

In the production agriculture workshop about a third of the recommended energy conservation projects dealt with systems studies and management practices. One study that will have an impact on all the others is determination of the effect of continuous residue removal on subsequent crop production.

Many proposals to utilize plant biomass as an energy resource are based on the premise that it is unnecessary to return all residue to the land. The exact amount of surplus organic matter available must be known before biomass energy schemes can be undertaken.

TABLE 97. – RECOMMENDED ENERGY CONSERVATION PROJECTS
FOR PRODUCTION AGRICULTURE

Crop production systems

1. Reduction of transport energy in crop production
2. Energy conservation through reduced tillage
3. Cropping for more energy-efficient food chains
4. Reduction of harvesting losses
5. Multiple cropping for energy efficiency
6. Reduced vulnerability to weather
7. Alternative portable fuel systems
8. Systems approach to energy conservation
9. Effect of continuous crop residue removal on soil productivity
10. Agricultural utilization of nonagricultural by-products
11. Energy conservation through genetic improvement
12. More efficient nutrient utilization
13. Energy reduction through increased pesticide efficiency
14. Biological nitrogen fixation

Animal production systems

15. Improvement of the energy efficiency of animals by physiological and nutritional means
16. Matching of animal production to processing or manufacturing process
17. Recovery and development of energy from animal excreta
18. Conservation of energy used for handling livestock waste
19. Conservation of energy in the construction and use of buildings for livestock production
20. Conservation of energy used to control the environment in livestock housing
21. Improved motor efficiency and more efficient use of motors in farmstead equipment
22. Energy conservation in feed-handling equipment on the farmstead
23. Minimizing energy costs of environmental modifications in animal production systems
24. Energy self-sufficient animal production systems
25. Future requirements for energy in animal and poultry production
26. Conservation of energy in transport
27. Integrated milk production and sanitation systems
28. Utilization of industrial low-temperature waste heat
29. Development and demonstration of forage and grain handling systems
30. Promotion, acceptance, and adoption of energy-saving methods of livestock production
31. Potential for energy-flexible systems

TABLE 97. — RECOMMENDED ENERGY CONSERVATION PROJECTS
FOR PRODUCTION AGRICULTURE *(continued)*

32. Study of meat animals and energy use
33. Development of effective animal manure management systems
34. Improved milk handling procedures

Greenhouse and other production systems

35. Energy-effective use of lighting systems for heating and constant year-round production
36. Reduced heat losses in greenhouses
37. Reduced energy consumption for cooling greenhouses
38. Improved greenhouse heating systems
39. Greenhouse orientation and shape
40. Increased space utilization in greenhouses
41. Development of new construction materials for greenhouses
42. Biological and physiological alternations of crops
43. Systems analysis of intensive vegetable, grain and flower production for energy conservation
44. Reduction of energy use in the transport and marketing of vegetables, fruits and flowers
45. Substitution of scarce fuels
46. Combination systems for greenhouse production
47. Increased efficiency of materials handling systems in aquacultural production
48. Utilization of low-grade waste heat in aquacultural production
49. More efficient vehicle use in aquacultural production
50. Domestic aquacultural production
51. Combination systems for aquacultural production
52. Establishment of a data base for energy conservation in controlled-environment agriculture

On-the-farm processing

53. Annual cycle of energy systems applied to farm processes
54. Milk volume reduction systems
55. Demonstration of energy conservation through the recycling of heat from milk refrigeration units
56. Energy conserved and effect on electrical demands resulting from shifting peak load
57. Establishing energy use patterns for livestock feed processing and handling systems
58. Grain-drier fuel from non-grain plant parts
59. Heat recovery systems for grain driers
60. Low-temperature grain drying
61. Innovative design for grain driers
62. Investigation of the technical and economic feasibility of a coal-fired grain drier
63. Evaluation of technical and economic feasibility of heat pumps for grain drying
64. Direct oil-fired burner for crop drying
65. Utilization of lined tunnels below grade as a heat exchanger
66. Preservatives to delay drying as applied to cereal grains
67. Development and demonstration of "dryeration"
68. Use of industrial waste heat for crop drying
69. Containerization for low-energy grain handling

TABLE 97. — RECOMMENDED ENERGY CONSERVATION PROJECTS
FOR PRODUCTION AGRICULTURE *(concluded)*

Water resources

70. National impact of energy conservation on water resources
71. Energy limitations for the reclamation of submerged lands
72. Irrigation management
73. Energy conservation due to improved pumping-plant efficiencies
74. Development of educational material for evaluating alternatives to improved pumping-plant efficiency
75. Alternative sources of energy for pumping water: (*a*) wind; (*b*) solar energy
76. Improvement of motor and engine efficiencies
77. Improvement of pump efficiency
78. Development of new well screens and gravel-pack procedures to reduce drawdown in wells
79. Irrigation system efficiency
80. Water application efficiency
81. Crop response to limited water and nitrogen inputs
82. Management of pumped water
83. Energy conservation in agricultural drainage installations
84. Energy conservation through improved drainage pumping efficiencies
85. Drainage materials and design practices for energy conservation
86. Development of a simplified flow meter
87. Measures for retaining water on land to reduce irrigation pumping
88. Water harvesting as an alternative to pump irrigation
89. Recharging of aquifers to reduce lift in pump irrigation

Utilization of agricultural by-products

90. Development of methods for harvesting, storing and transporting crop residues
91. Availability of field crop residues
92. Definition of economical methods of whole crop harvesting for optimum grain quality, energy use, and residue collection
93. Practicality of redesigning grain crops
94. Demonstration of direct combustion of field crop residues
95. Off-the-farm use of field crop residues
96. Gas generators for farm and industrial use
97. Definition and demonstration of optimum liquid-fuel production systems
98. Definition and demonstration of optimum systems for the production of industrial chemicals from agricultural residues
99. Increased developmental efforts on the enzymatic hydrolysis of cellulosic residues to sugars
100. Demonstration of practical methods of CO_2 enrichment
101. Integration of feeding and fuel production in the use of crop residues
102. Use of the mineral content of crop residues
103. Relative efficiency and economic aspects of utilizing manure
104. Development of energy-efficient equipment to process livestock wastes for refeeding
105. Health and consumer considerations in feeding animal wastes to livestock
106. Demonstration of cascade feeding systems for poultry, steer and brood cattle
107. Feasibility of converting feedlot wastes into industrially useful oxychemicals

SOURCE: (71)

Chemicals

1. Alternative sources of feedstocks for ammonia plants
2. Use of coal as a fuel for ammonia plants rather than natural gas
3. Conservation of energy in the production of phosphoric acid
4. Comparative study of total energy consumption of marketing systems for anhydrous ammonia versus nitrogen solution
5. Conservation of energy in granulation plants
6. Improved head recovery in existing ammonia plants
7. Conservation of energy in phosphate rock mining by extracting P_2O_5 directly from rock-clay matrix
8. Combined operation of an anhydrous ammonia plant and a steel blast furnace
9. Conservation of energy through minimum tillage
10. More efficient pollution-control equipment in fertilizer plants
11. Conservation of electric power in granulation plants
12. Efficient marketing and distribution of fertilizers and agricultural chemicals
13. Use of by-products in agriculture
14. Solar evaporation of potash brines to concentrate and crystallize potash
15. Wet-grinding of phosphate rock

Dairy

16. Development and dissemination of technical educational materials and of technology transfers
17. Development of alternatives to current package-sealing techniques
18. Integrated energy systems for dried food products
19. Utilization of heat pumps in manufacturing processes
20. Utilization of solid wastes as an energy source
21. Development of 2:1 sterile concentrate milk beverage
22. Identification and evaluation of constraints on energy use in the dairy products industry
23. Improvement in the energy efficiency of plant clean-up practices
24. Feasibility study of sterile fluid milk as a major conservation measure; development and demonstration of a pilot system

Fruits and vegetables

25. Heat recovery: gases and water
26. Methane production and anaerobic digestion of wastes
27. Improved steam traps
28. "Good management"
29. Systems studies of industrial parks
30. Fuel substitution
31. Improved combustion efficiency
32. Improved refrigeration equipment and processes
33. Reduction in bulk to decrease handling, warehousing and transport costs
34. Earth-covered and underground storage of cold and frozen products
35. Packaging
36. Water removal
37. Sterilization
38. Equipment ratings
39. Evaluation of current practices in food-processing firms
40. Education of the public

TABLE 98. — RECOMMENDED ENERGY CONSERVATION PROJECTS
FOR THE AGRICULTURAL PROCESSING INDUSTRY *(continued)*

Grain

41. Improved heat-transfer and heat-recovery capability in drying systems
42. Improved technology for process and moisture control
43. Liquid concentration
44. Improved heat-recovery systems
45. Alternative fuels to replace natural gas
46. Alternative processing and formulation methods to reduce energy consumption
47. Improved equipment design

Meat

48. Systems analysis of energy use in the meat industry
49. Automatic air/fuel ratio control in small boilers
50. Recycling of waste water effluents through a reclamation system to produce crystal-clear potable water
51. Re-use of hot and/or tempered water through filtration
52. Reduced energy use in beef and hog viscera inspection tables
53. Reduced energy use in high-pressure and high-temperature beef carcass washers
54. Substitution of sanitation for refrigeration to reduce energy used in chilling red meats
55. Elimination of after-burners
56. Use of outdoor winter air for refrigeration in cold climates
57. Recovery of escaping heat in waste effluent
58. Reduced natural gas consumption in hog singeing
59. Development of less energy-intensive means of eliminating condensation on plant ceilings
60. Hot deboning
61. Industrial boiler systems
62. Minimization of packaging material for meats
63. Development of alternative methods for chilling carcasses
64. Development of a new method of removing water from material that is to be rendered inedible
65. Better ammonia and Freon refrigeration systems and controls for defrosting
66. Raising the gauge of low-pressure steam by compression
67. Elimination of steam traps by introducing a sump at a lower level and pumping condensate under pressure to steam generator
68. Improved scalding and de-hairing methods for hogs with mechanical equipment and chemicals
69. Microwave processing
70. Pressurized meat-processing smokehouses
71. Reduced energy use in refrigerated transport
72. Demonstration of energy conservation measures
73. Vacuum chilling and storage of red meats
74. Immersion chilling of red meats
75. Reduction of the amount of water used in all meat- and poultry-processing plants
76. Studies of the effects of a four-day, forty-hour work week on energy consumption and on the economies of various plant capacities

Textiles

77. Conversion from pneumatic to mechanical materials handling at gins
78. Use of ginning waste as a drying fuel

TABLE 98. — RECOMMENDED ENERGY CONSERVATION PROJECTS
FOR THE AGRICULTURAL PROCESSING INDUSTRY *(concluded)*

79. Use of ginning waste in the generation of power for cotton gins
80. Compression process using hydraulic power instead of steam power
81. Reduction of energy consumption in wool scouring
82. Reduction of hot water use in wet-processing of textiles containing cotton
83. Elimination of drying processes in wet-processing of textiles containing cotton
84. New fibre-to-yarn techniques for agricultural fibres
85. Field cleaning of seed cotton
86. Improved lint cleaning at gins
87. More efficient gin seed cotton driers

SOURCE: (70)

A total systems study of agricultural production was recommended to determine interactions and interdependencies for the purpose of finding possible energy trade-offs. In addition, microenvironmental effects on energy use in animal production should be determined. Improving the efficiency of animals by physiological and nutritional means affords potential energy savings.

The book *Energy for Rural Development* (4) outlines various energy resources, such as direct solar heating and cooling, photosynthetic products in the form of organic materials, direct generation of electricity, wind energy, hydropower and geothermal energy. The current state of these alternatives is presented along with five- and ten-year forecasts of the technology likely to become available for their application in rural areas of the developing countries. The following sections summarize the findings of the National Academy of Sciences energy panel. ·

Developing new energy sources

DIRECT SOLAR HEATING AND COOLING

One of the agricultural uses of solar energy is crop drying both by direct exposure to sunlight and by passing solar-heated air through or over the material. The cost of drying grains, fruits, lumber and other products with solar-heated air has not yet been reliably appraised. Although solar drying systems appear to be slightly more costly than conventional drying schemes, they are within economic reach.

Solar energy has been used successfully for domestic hot-water supplies. As such systems are often economically competitive with others, their economic advantage over systems using fossil fuels will increase as fuel prices rise. Where fuels such as wood or peat are locally available at little or no cost — although supplies are diminishing — the capital cost of solar-heated domestic hot-water supply may not be justified.

Residential heating by solar energy often compares unfavourably where fossil fuels are cheap and plentiful — a situation that appears to be vanishing rapidly. Hot-air systems, which have been used successfully for residential heating, are simple and reliable; moreover, they can often be realized with local materials, làbour and know-how.

The production of salt by evaporation of sea water or inland brines is a direct use of solar energy immediately available wherever water containing salt occurs in conjunction with an appropriate climate.

In the near future the developing countries should be able to initiate viable solar-energy applications. Some materials would have to be imported, but the greater part would be available in most countries. The availability of capital is the major hindrance to the near-future application of solar energy in the developing countries.

Within five to ten years the development of improved cooling systems can be expected. In compression-type systems, improvements and economies in the organic Rankine cycle and possibly in the Stirling cycle are likely. With regard to absorption processes, research should advance multistage systems and new working fluids.

In about ten years small solar-powered electrical generators may become practical — especially for communities remote from electric networks. In areas with very high fuel prices and abundant solar energy, this technology could be introduced if capital were available.

As for large-scale generation of electric power, little will be accomplished in the developing countries in the next ten years. This may be due less to the rate of technological progress than to lack of the highly complex industrial infrastructure and the considerable capital needed for successful implementation.

PHOTOVOLTAIC DEVICES

Direct conversion of the visible part of the solar spectrum — sunlight — to electricity is perhaps the neatest and most ecologically attractive of all schemes for the exploitation of solar energy. It is also, unfortunately, one of the least feasible for widespread application in the developing nations.

Direct photovoltaic conversion can be achieved with basically simple devices (solar cells) without moving parts that require no additional energy and little, if any, maintenance. A reduction in the cost of solar cells may make their use economically feasible and justifiable for rural areas. At present, the developing countries, with very few exceptions, can avail themselves of this technology only by importing the components.

Photovoltaic systems with 10% solar conversion efficiency and with peak power capacities ranging from one watt to tens of kilowatts are available from manufacturers in Japan, the U.S.A., the United Kingdom, France and the Federal Republic of Germany. The price of a complete system is determined by the photovoltaic array. Prices of $20 to $30 per peak watt, corresponding to $100 000 to $150 000 per average kilowatt installed capacity, are now being quoted; however, these prices are expected to decrease by a factor of four within the next few years.

In the coming five years over $100 million will be spent on the development of lower-cost terrestrial photovoltaic power systems. Arrays incorporating wide-aperture concentrators without diurnal tracking requirements will probably be on the market by 1980.

Commercial photovoltaic conversion arrays costing a few hundred dollars per average kilowatt are likely to become available by 1985 or sooner. The integration of these arrays into a complete system of energy storage, power conditioning, and transmission and distribution is necessary before they can be used on any substantial scale for power generation greater than a few kilowatts.

WIND ENERGY

There are enough commercially built windmills currently available to permit immediate application of wind energy, whether it be for pumping water, compressing air or generating electricity.

The water-pumping units vary significantly in price from $5 000 to $10 000 per kilowatt for fractional-size units to $1 300 to $2 600 per kilowatt for units of 7 to 20 kilowatts. Few larger pumping units are commercially available today.

Electric generating units cost from $3 000 to $6 000 per kilowatt for fractional-kilowatt generators or from $1 000 to $2 000 per kilowatt for units in the 5- to 15-kilowatt range. A unit that is going into production in the U.S.A. is expected to sell for about $500 per kilowatt. No larger commercial units are being marketed at present.

Most commercially available windmill units sold today are safe and dependable, require little maintenance and are built to last for ten to twenty years (sometimes more) with proper maintenance.

In developing areas, classical windmills often cost much less because of local technologies.

HYDROPOWER

Small hydroelectric generators are available from manufacturers in the U.S.A. and Europe. Because of the limited production, the cost per kilowatt is quite high. Units capable of generating 10 kilowatts cost about $1 000 per kilowatt. The cost per kilowatt increases with smaller-capacity units.

These costs do not include the other necessary components of the system, such as the dam, the piping connection between the dam and the turbine, and the housing. Considering all outlays, the cost of electricity at the generator will probably range from $0.08 to $0.10 per kilowatt-hour.

PHOTOSYNTHETIC PRODUCTS AS AN ENERGY SOURCE

The one renewable source on which mankind has relied since the discovery of fire is photosynthesis. Even though nature performs the initial function of converting solar energy into energy-rich biomass, the photosynthetic process is subject to technological advances as sophisticated in their own way as those of photovoltaics.

When human utilization of sunlight is viewed as an energy source, the criteria for choosing and timing crops differ from those conventionally applied, as factors commonly ignored then become very significant. The energy content of the plant material is especially important where there is a choice among food or fibre crops with different amounts of residue containing exploitable energy (202).

Furthermore, for maximum use of available sunlight the farmland must be kept covered with green plants as much of the year as possible. This involves, for example, multiple cropping, intercropping, nurse crops (i.e., later-maturing crops planted with early-maturing crops, so that the former will have a substantial start when the latter are harvested), and the cultivation of cool-, warm-, wet-, and dry-season plants in those periods when traditional crops cannot be grown.

Most agricultural schemes convert into plant material less than 1% of the solar energy available during a growing season. Some plants, however, have a more efficient photosynthetic system and can convert 2–3% of the incident solar energy. Known as C_4 plants, they require warm growing seasons and high levels of solar radiation. They are already widely grown in many of the developing countries and warrant serious consideration in other countries with appropriate climates where they are not grown. As these species use water more efficiently, they are more tolerant of drought than many common crop species.

Another group known as CAM plants, owing to a genetically controlled metabolic characteristic, are among the world's most efficient water users and therefore do well in semi-arid regions. They are slow-growing plants, but in dry areas, where plants can be harvested on long-term rotation (2–6 years), they would make the land more productive.

ALCOHOL PRODUCTION

The chemical production of ethyl alcohol by fermentation is similar to the microbial production of acetone, butyl alcohol and isopropyl alcohol — all flammable liquids. Before fuel production utilizing microbiological processes can become realistic in rural areas of developing nations, research and development are needed on materials, equipment and processing. To reduce capital investment and the need for advanced technical skills, it will be necessary to explore the local availability of raw materials, to investigate the microbiology and physical chemistry of the fermentation systems and to examine recovery operations and equipment.

GEOTHERMAL ENERGY

In areas of the world that have experienced volcanic activity, geothermal resources that can be exploited economically may exist. In the Philippines, for example, geothermal energy might be the most promising indigenous resource, and its development would greatly reduce dependence on oil (130).

Geothermal energy provides a source of energy for electricity generation, space heating, drying and refrigeration that is competitive with other technologies. Technology for exploiting geothermal sources is in use in many countries. Geothermal turbine systems appear to have a life of about thirty years or more. The life of hot-water wells appears to be at least ten to fifteen years — longer if used with space-heating systems.

Costs range from $150 to $500 per kilowatt capacity for power plants and from about $70 to $100 per kilowatt for wells. A minimum exploration programme may cost about $100 000, but can easily reach $1 million if much exploratory drilling is required.

Geothermal power projects have proved to be 80–95% reliable. Hot-water systems and dry-steam systems are in operation in Japan, New Zealand, Mexico and the U.S.A.

A new device for hot-water or steam systems — the screw expander —

is being developed in the U.S.A. Two counter-rotating screws serve as a positive displacement device within which the hot geothermal water boils and the steam expands, thereby rotating the screws. The screw expander provides mechanical energy or, in conjunction with a generator, electric power. A test programme with a 1.25-megawatt device, sponsored by the National Science Foundation in the U.S.A., is under way at the Jet Propulsion Laboratory of the California Institute of Technology.

PETROLEUM PRODUCTS

There is cause for concern about future supplies and prices of petroleum products, and by no means should the importance of petroleum be minimized. No one knows how much petroleum remains to be discovered or can be profitably extracted from the ground by the new oil-recovery technology prompted by higher oil prices. There is no really suitable alternative to petroleum for the next two decades, beyond which future energy technology is highly speculative. Therefore, wise use by all the nations of the world is essential. International efforts to conserve petroleum and eliminate waste — to stretch out the use of the precious supply of petroleum for as long as possible — should be a matter of top priority.

It can be argued that even poor countries can afford to purchase petroleum for food production in the foreseeable future. Revelle (201) reasons that if the need for additional energy for draught power, water pumping and chemical fertilizer manufacture in India were to be met by petroleum products, 43 million metric tons would be needed. If all this petroleum were imported, the cost could be defrayed by the export of about one tenth of the increased foodgrain production; but the crop residues from 300 million metric tons of foodgrains would be at least 400 million metric tons, with an energy content of 13×10^{14} kilocalories, about 2.7 times the required additional energy. If this energy provided by photosynthesis could be harnessed — for example, in nitrogen-conserving fermentation plants of the type suggested by Makhijani and Poole (151) — the Indian rural ecosystem would continue to be fairly self-sufficient.

Ultimately, petroleum resources will be exhausted and alternative energy resources must be developed. Great emphasis should therefore be placed on developing alternatives, such as solar energy, breeder reactors and fusion. Our future rests on our ability to shift to a new energy base.

What should be done when the world's supply of petroleum becomes insufficient to satisfy all the needs of mankind? Of course, energy prices can be expected to increase; however, price allocation alone may not be in the best interest of mankind, especially in the poorer countries. It is conceivable that priorities will have to be developed and mechanisms other than price established for allocating scarce petroleum supplies. Should this point be reached, it seems that agriculture should be given top priority.

UNITS OF MEASURE AND CONVERSION FACTORS

Most measured quantities can be described in terms of three basic dimensions: length, mass and time. To determine speed, for example, length is divided by time. The common units of measurement are listed below.

Length (L)	Area (L²)	Volume (L³)
mile (mi)	square mile (mi²)	gallon (gal)
yard (yd)	square yard (yd²)	quart (qt)
foot (ft)	square foot (ft²)	pint (pt)
inch (in)	square inch (in²)	ounce (oz)
fathom (fath)	acre	cubic foot (ft³)
kilometre (km)	square kilometre (km²)	cubic yard (yd³)
metre (m)	square metre (m²)	cubic inch (in³)
centimetre (cm)	square centimetre (cm²)	litre (l)
micron (μ)		cubic centimetre (cm³)
ångström (Å)		acre-foot
		cord (cd)
		cord-foot
		barrel (bbl)

Mass (M)	Speed (L/T)	Flow rate (L³/T)
pound (lb)	feet/minute	cubic feet/minute
ton (short)	feet/second	cubic metres/minute
ton (long)	miles/hour	litres/second
ton (metric)	miles/minute	gallons/minute
gram (g)	kilometres/hour	gallons/second
kilogram (kg)	kilometres/minute	
	kilometres/second	

Pressure (M/L/T²)	Energy (ML²/T²)	Power (ML²/T³)
atmosphere (atm)	British thermal unit (Btu)	British thermal units/minute

Pressure (M/L/T²)	Energy (ML²/T²)	Power (ML²/T³)
pounds-force/ square inch (lbf/in²)	calorie (cal)	British thermal units/hour
inch of mercury (Hg)	foot-pound	watt
centimetre of mercury (Hg)	joule	joules/second
feet of water	kilowatt-hour (kWh)	calories/minute
	horsepower-hour (HPh)	horsepower (HP)

Time (T)	Energy density (M/T²)	Power density (M/T³)
year	calories/square centimetre	calories/square centimetre/ minute
month	British thermal units/square foot	British thermal units/square foot/hour
day	langley	langleys/minute
hour (h)	watt-hours/square foot	watts/square metre
minute (min)		
second (s)		

The factors used to convert one set of units to another are given below. It may be necessary to use the reciprocal of the conversion factor or to make more than one conversion to obtain the units required.

Multiply	By	To obtain
acres	43560	square feet
	0.004047	square kilometres
	4047	square metres
	0.0015625	square miles
	4840	square yards
acre-feet	43560	cubic feet
	1233.5	cubic metres
	1613.3	cubic yards
ångströms	1×10^{-8}	centimetres
	3.937×10^{-9}	inches
	0.0001	microns
atmospheres	76	centimetres of Hg (0°C)

Multiply	By	To obtain
atmospheres	1033.3	centimetres of H_2O (4°C)
	33.8995	feet of H_2O (39.2°F)
	29.92	inches of Hg (32°F)
	14.696	pounds-force/square inch
barrels (petroleum, U.S.)	5.6146	cubic feet
	35	gallons (Imperial)
	42	gallons (U.S.)
	158.98	litres
British thermal units	251.99	calories, grams
	777.649	foot-pounds
	0.00039275	horsepower-hours
	1054.35	joules
	0.000292875	kilowatt-hours
	1054.35	watt-seconds
British thermal units/hour	4.2	calories/minute
	777.65	foot-pounds/hour
	0.0003927	horsepower
	0.000292875	kilowatts
	0.292875	watts (or joules/second)
British thermal units/pound	7.25×10^{-4}	calories/gram
British thermal units/square foot	0.271246	calories/square centimetre (or langleys)
	0.292875	watt-hours/square foot
British thermal units/square foot/hour	3.15×10^{-7}	kilowatts/square metre
	4.51×10^{-3}	calories/square centimetre/minute (or langleys/minute)
	3.15×10^{-8}	watts/square centimetre
calories	0.003968	British thermal units
	3.08596	foot-pounds
	1.55857×10^{-6}	horsepower-hours
	4.184	joules (or watt-seconds)

Multiply	By	To obtain
calories	1.1622×10^{-6}	kilowatt-hours
calories (food unit)	1 000	calories
calories/minute	0.003968	British thermal units/minute
	0.06973	watts
calories/square centimetre	3.68669	British thermal units/square foot
	1.0797	watt-hours/square foot
calories/square centimetre/minute	796320	British thermal units/square foot/hour
	251.04	watts/square centimetre
candle power (spherical)	12.566	lumens
centimetres	0.032808	feet
	0.3937	inches
	0.01	metres
	10 000	microns
centimetres of Hg (0°C)	0.0131579	atmospheres
	0.44605	feet of H_2O (4°C)
	0.19337	pounds/square inch
centimetres of H_2O (4°C)	0.0009678	atmospheres
	0.01422	pounds/square inch
centimetres/second	0.032808	feet/second
	0.022369	miles/hour
cords	8	cord-feet
	128 (or $4 \times 4 \times 8$)	cubic feet
cubic centimetres	3.5314667	cubic feet
	0.06102	cubic inches
	1×10^{-6}	cubic metres
	0.001	litres
	0.0338	ounces (U.S., fluid)
cubic feet	0.02831685	cubic metres
	7.4805	gallons (U.S., liquid)
	28.31685	litres
	29.922	quarts (U.S., liquid)
cubic feet of H_2O (60°F)	62.366	pounds of H_2O
cubic feet/minute	471.947	cubic centimetres/second
cubic inches	16.387	cubic centimetres

Multiply	By	To obtain
cubic inches	0.0005787	cubic feet
	0.004329	gallons (U.S., liquid)
	0.5541	ounces (U.S., fluid)
cubic metres	1×10^6	cubic centimetres
	35.314667	cubic feet
	264.172	gallons (U.S., liquid)
	1000	litres
cubic yards	27	cubic feet
	0.76455	cubic metres
	201.97	gallons (U.S., liquid)
cubits	18	inches
fathoms	6	feet
	1.8288	metres
feet	30.48	centimetres
	12	inches
	0.00018939	miles (statute)
feet of H_2O (4°C)	0.029499	atmospheres
	2.2419	centimetres of Hg (0°C)
	0.433515	pounds/square inch
feet/minute	0.508	centimetres/second
	0.018288	kilometres/hour
	0.0113636	miles/hour
foot-candles	1	lumens/square foot
foot-pounds	0.001285	British thermal units
	0.324048	calories
	5.0505×10^{-7}	horsepower-hours
	3.76616×10^{-7}	kilowatt-hours
furlong	220	yards
gallons (U.S., dry)	1.163647	gallons (U.S., liquid)
gallons (U.S., liq)	3785.4	cubic centimetres
	0.13368	cubic feet
	231	cubic inches
	0.0037854	cubic metres
	3.7854	litres
	8	pints (U.S., liquid)
	4	quarts (U.S., liquid)
gallons/minute	2.228×10^{-3}	cubic feet/second
	0.06308	litres/second
grams	0.035274	ounces (avoirdupois)
	0.002205	pounds (avoirdupois)
gram-centimetres	9.3011×10^{-8}	British thermal units
grams/square metre	3.98	short tons/acre
	8.92	pounds/acre

Multiply	By	To obtain
horsepower	42.4356	British thermal units/minute
	550	foot-pounds/second
	745.7	watts
horsepower-hours	2546.14	British thermal units
	641616	calories
	1.98×10^6	foot-pounds
	0.7457	kilowatt-hours
inches	2.54	centimetres
	0.83333	feet
inches of Hg (32°F)	0.03342	atmospheres
	1.133	feet of H_2O
	0.4912	pounds/square inch
inches of H_2O (4°C)	0.002458	atmospheres
	0.07355	inches of Hg (32°F)
	0.03613	pounds/square inch
joules	0.0009485	British thermal units
	0.73756	foot-pounds
	0.0002778	watt-hours
	1	watt-seconds
kilocalories/gram	1378.54	British thermal units/pound
kilograms	2.2046	pounds (avoirdupois)
kilograms/hectare	0.893	pounds/acre
	0.0004465	short tons/acre
kilometres	1000	metres
	0.62137	miles (statute)
kilometres/hour	54.68	feet/minute
kilowatts	3414.43	British thermal units/hour
	737.56	foot-pounds/second
	1.34102	horsepower
kilowatt-hours	3414.43	British thermal units
	1.34102	horsepower-hours
knots	51.44	centimetres/second
	1	miles (nautical)/hour
	1.15078	miles (statute)/hour
langleys	1	calories/square centimetre
litres	1000	cubic centimetres
	0.0353	cubic feet
	0.2642	gallons (U.S., liquid)
	1.0567	quarts (U.S., liquid)

Multiply	By	To obtain
pounds/acre	0.0005	short tons/acre
litres/minute	0.0353	cubic feet/minute
	0.2642	gallons (U.S., liquid)/minute
lumens	0.079577	candle power (spherical)
lumens (at 5 550 A)	0.0014706	watts
metres	3.2808	feet
	39.37	inches
	1.0936	yards
metres/second	2.24	miles/hour
microns	10000	ångströms
	0.0001	centimetres
miles (statute)	5280	feet
	1.6093	kilometres
	1760	yards
miles/hour	44.704	centimetres/second
	88	feet/minute
	1.6093	kilometres/hour
	0.447	metres/second
millilitres	1	cubic centimetres
millimetres	0.1	centimetres
ounces (avoirdupois)	0.0625	pounds (avoirdupois)
ounces (U.S., liquid)	29.57	cubic centimetres
	1.8047	cubic inches
	0.0625 (1/16)	pint (U.S., liquid)
pints (U.S., liquid)	473.18	cubic centimetres
	28.875	cubic inches
	0.5	quarts (U.S., liquid)
pounds (avoirdupois)	0.45359	kilograms
	16	ounces (avoirdupois)
pounds of water	0.01602	cubic feet of water
	0.1198	gallons (U.S., liquid)
pounds/acre	0.0005	short tons/acre
pounds/square inch	0.06805	atmospheres
	5.1715	centimetres of Hg (0°C)
	27.6807	inches of H_2O (39.2°F)
quarts (U.S., liquid)	0.25	gallons (U.S., liquid)
	0.9463	litres
	32	ounces (U.S., liquid)
	2	pints (U.S., liquid)
radians	57.30	degrees

Multiply	By	To obtain
square centimetres	0.0010764	square feet
	0.1550	square inches
square feet	2.2957×10^{-5}	acres
	0.09290	square metres
square inches	6.4516	square centimetres
	0.006944	square feet
square kilometres	247.1	acres
	1.0764×10^7	square feet
	0.3861	square miles
square metres	10.7639	square feet
	1.196	square yards
square miles	640	acres
	2.788×10^7	square feet
	2.590	square kilometres
square yards	9 (or 3×3)	square feet
	0.83613	square metres
tons (long)	1016	kilograms
	2240	pounds (avoirdupois)
tons (metric)	1000	kilograms
	2204.6	pounds (avoirdupois)
tons (metric)/ hectare	0.446	short tons/acre
tons (short)	907.2	kilograms
	2000	pounds (avoirdupois)
watts	3.4144	British thermal units/hour
	0.05691	British thermal units/minute
	14.34	calories/minute
	0.001341	horsepower
	1	joules/second
watts/square centimetre	3172	British thermal units/ square foot/hour
watt-hours	3.4144	British thermal units
	860.4	calories
	0.001341	horsepower-hours
yards	3	feet
	0.9144	metres

ENERGY CONVERSION UNITS SHOWN
ON A LOGARITHMIC SCALE

Note that a kilowatt-hour is 3.6 million times greater than a joule.

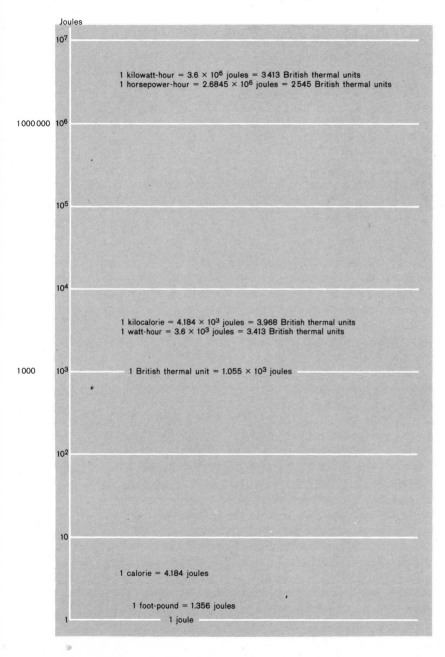

REFERENCES

1. ABELSON, PHILIP H. Energy from biomass. *Science*, 191 (4233): 1.
   1976

2. ABERNATHY, G.H. & COOKE, M.D. JR. *Factors affecting the efficiency of natural*
   1977 *gas powered pumping plants*. Las Cruces, New Mexico, New Mexico
   State University. ASAE Paper No. 77-2017. 9 p.

3. ABERNATHY, G.H. & MANCINI, T.R. *Design and installation of a solar-powered*
   1977 *irrigation pump*. Las Cruces, New Mexico, New Mexico State University.
   ASAE Paper No. 77-4020. 7 p.

4. AD HOC PANEL OF THE ADVISORY COMMITTEE ON TECHNOLOGY. *Energy for*
   1976 *rural development*. Washington, D.C., Board of Science and Technology
   for International Development and Commission on International Relations, National Academy of Sciences.

5. ALDER, E.F., KLINGMAN, G.C. & WRIGHT, W.L. Herbicides in the energy
   1976 equations. *Weed Science*, 24 (1): 99-106.

6. ALLAWAY, W.H. *The effect of soils and fertilizers on human and animal nu-*
   1975 *trition*. Washington, D.C., U.S. Department of Agriculture, Agricultural
   Research Service. Agriculture Information Bulletin No. 378. 52 p.

7. AMERICAN PHYSICAL SOCIETY. Report to APS by the Study Group on Light
   1975 Water Reactor Safety. Presented at 1975 Spring Meeting of the Society.
   Reviews of Modern Physics, Vol. 14, Supplement 1.

8. ANDERSON, L.L. *Energy potential from organic wastes: a review of the quantities*
   1972 *and sources*. Washington, D.C., U.S. Department of Interior, Bureau
   of Mines. Information Circular 8549. 14 p.

9. ANDROSKY, A. *Uses of the sun in the service of man*. El Segundo, California,
   1973 Aerospace Corporation. Report No. ATR-74-(9470). 35 p.

10. ARMSTRONG, C.H. *Technology advances raising combined cycle efficiency*. Paper
    1974 presented at the Gas Turbine International.

11. ASIMOV, I. *Worlds within worlds: the story of nuclear energy*. Vol. 3. *Nuclear*
    1972 *fission, nuclear fusion, beyond fusion*, p. 113-164. Washington, D.C.
    ERDA, Office of Public Affairs.

12. ATOMIC ENERGY COMMISSION. *Reactor safety study: an assessment of accident*
    1974 *risks in U.S. commercial power plants*. Washington, D.C. WASH-1400
    Draft.

13. Atomic Energy Research Establishment, Harwell. Electronics and Applied Physics Division. Oxfordshire, United Kingdom.

14. BACHER, KEN. *Putting the sun to work.* Phoenix, Arizona, Arizona State 1974 Fuel and Energy Office. 27 p.

15. BARGER, E.L. *Tractors and their power units.* New York, Wiley. 496 p. 1963

16. BATTELLE SEATTLE RESEARCH CENTER. *Geothermal energy.* Seattle, Washington. 1972 77 p.

17. BERTIN ET AL. *Production d'alcool d'origine végétale.* 1975

18. BLANC, M. *L'utilisation de l'énergie éolienne pour le pompage de l'eau.* BCEOM-1975 Information et documents.

19. BLOUIN, G.M. *Effects of increased energy costs on fertilizer production costs* 1974 *and technology.* Knoxville, Tennessee, Tennessee Valley Authority. Bulletin Y-84. 30 p.

20. BOGUE, D.J. *Principles of demography.* New York, Wiley. 1969

21. BOHR, K. & SADOVE, R. *Some guides to forecasting electric power demand* 1958 *in underdeveloped areas.* Washington, D.C., International Bank for Reconstruction and Development.

22. BONISCENTI, M.V. Economie énergétique dans le secteur des tracteurs. *Mé-1975 canisation et problèmes techniques, 16è réunion, Bologne, Italie, avril 29-30, 1975.*

23. BP COMPANY LIMITED. *BP statistical review of the world oil industry, 1975.* 1976 32 p.

24. BRAQUAVAL, R. Les diverses sources d'énergie: leurs applications aux pays 1974 en voie de développement. *Industries et travaux d'outre-mer,* décembre, p. 1051-1057.

25. BRELLE, Y. ET AL. *Principes, technologie, applications des piles à combustibles.* 1972

25a. BROOKHAVEN NATIONAL LABORATORY DEVELOPING COUNTRIES ENERGY PRO-1978 GRAM. *Energy needs, uses and resources in developing countries.* BNL 50784. March.

26. BROWN, H. Energy in our future. *In* Hollander, J.M. and Simmons, M.K. 1976 *Annual Review of Energy,* Vol. 1, p. 1-36. Palo Alto, California, Annual Reviews. 793 p.

27. BURNS, R.C. & HARDY, R.W.F. *Nitrogen fixation in bacteria and higher plants.* 1975 New York, Springer-Verlag.

28. *Business Week.* U.S. power ultimate weapon in world politics. December, 1975 1975.

29. BUTCHER, WALTER, HINMAN, G., LEOPOLD, A.C., PIMENTEL, D., ROSS, J. &
    1975 YOUNG, E.J. Energy aspects of world food problems. *Energy Com-*
        mission of the Institute of Ecology, Founders' Conference, Morgantown,
        W. Va, April 16-19, 1975. 23 p.

30. BUTLER, J.L. & TROEGER, J.M. *Use of solar energy to cure peanuts.* Tifton,
    1977 Georgia, Coastal Plain Experiment Station. ASAE Paper No. 77-3033.

31. CENTRE TECHNIQUE DU BOIS. *Le chauffage au bois.* Série 1. *Connaissance*
    1975 *du bois. Traitements physiques et chimiques, préservation.* 40 p.

32. CALDERWOOD, D.L. *Rice drying with solar heat.* Beaumont, Texas, Agricultural
    1977 Research Service, U.S. Department of Agriculture. ASAE Paper No.
        77-3003. 11 p.

33. CAMPANA, ROBERT J. & LANGER, S. *Nuclear power and the environment.*
    1974 Hinsdale, Ill., American Nuclear Society. 64 p.

34. CARILLON, R. *Comment économiser le carburant et le combustible dans les*
    1975 *travaux agricoles (consommation directe).* Boulogne, Confédération
        européenne de l'agriculture.

35. CARILLON, R. *L'activité agricole et l'énergie.* Antony, Centre national d'études
    1975 et d'expérimentation de machinisme agricole. Etudes du CNEEMA,
        No. 408. 71 p.

36. CENTRE D'ÉTUDES ET D'EXPÉRIMENTATION DU MACHINISME AGRICOLE TROPICAL.
    1972 *Machinisme agricole tropical*, No. 38. 56 p.

37. CENTRE D'ÉTUDES ET D'EXPÉRIMENTATION DU MACHINISME AGRICOLE TROPICAL.
    1972 *The employment of draught animals in agriculture.* FAO. 249 p.

38. CENTRE NATIONAL D'ÉTUDES ET D'EXPÉRIMENTATION DE MACHINISME AGRICOLE.
    1974 *L'agriculture et les "carburants."* Antony, Etudes du CNEEMA, No.
        391. 61 p.

39. COLLI, JEAN-CLAUDE. *L'énergie solaire. Energies nouvelles.* Actualités-docu-
    1976 ments, Premier Ministre, Service d'information et de diffusion, N⁰ 11.
        Mars.

40. COMITÉ PROFESSIONEL DU PÉTROLE. *Pétrole.* 375 p.
    1975

41. Commission Federal de Electricidad, Mexico.

42. COMMONER, BARRY. *The poverty of power.* New York, Knopf.
    1976

43. CONGRÈS 1976 DE L'UPDEA. *Les microcentrales hydrauliques groupes bulbes.*
    1976 Communication de l'Electricité de France, Direction des affaires exte-
        rieures et de la coopération, p. 32-50.

44. CONNOR, L.J. *Beef feedlot design and management in Michigan.* East Lansing,
    1976 Michigan Agricultural Experiment Station. Report 292.

45. CONVERSE, J.C. & GRAVES, R.E. *Facts on methane production from animal*
    1974 *manure.* Wisconsin Extension Fact Sheet A 2636. 4 p.

46. Cook, Earl. The flow of energy in an industrial society. *Scientific American,*
    1974   225(3): 135-144.

47. Corliss, William, R. *Direct conversion of energy.* Oak Ridge, Tenn., U.S.
    1964   Atomic Energy Commission. Technical Information Center. 34 p.

48. Costes, C. et al. *Photosynthèse et production végétale.* Gauthiers-Villars.
    1975

49. Cottrell, Fred. *Energy and society: the relationship between energy, social*
    1974   *change and economic development,* p. 17-18.

50. Couston, J.W. Food and Agriculture Organization. Rome. April, 1974.

51. Cowell, Peter. Personal communication. Discussion of solar-powered water
    1976   pumps. Asia Institute of Technology, Bangkok, Thailand.

52. Crawley, Gerard, M. *Energy.* New York, Macmillan. 337 p.
    1975

53. Crosson, Pierre. *Interdependencies between food sector and non-food sector*
    1976   *activities with respect to energy resources and environment.* Washington,
    D.C. National Academy of Sciences. World Food and Nutrition
    Study, Sub-group 10B. 30 p.

54. Cunningham, J.P. *An energetic model linking forest industry and ecosystems.*
    1975   Helsinki, Finnish Forest Research Institute.

55. Daniels, F. *Direct use of the sun's energy.* New Haven, Conn., Yale Uni-
    1964   versity Press. 374 p.

56. Delwicke, C.C. The nitrogen cycle. *Scientific American,* September 1970:
    1970   137-146.

57. Duckert, Joseph M. *High-level radioactive waste: safe storage and ultimate*
    1975   *disposal.* Oak Ridge, Tenn., U.S. Atomic Energy Commission. 23 p.

58. Duff, Bart, Nichols, F.E., Campbell, J. & Lee, C.C. Agricultural equipment
    1974   development research for tropical rice cultivation. Los Baños, Agri-
    cultural Engineering Department, International Rice Research Institute.
    Semiannual Progress Report, p. 17-18.

59. Duncan, A. *Economic aspects of the use of organic matter as fertilizer.* FAO/
    1974   SIDA Expert Consultation on Organic Materials as Fertilizers, Rome,
    2-6 December 1974, p. 22. AGL/TMOF.

60. Dunn, P.D. Department of Engineering, University of Reading, U.K. Personal
    correspondence.

61. Durham, Tony. The breeder: fast and deadly. *Undercurrents,* 9: 17-32.
    1975

62. Duvant moteurs diesel. *Utilisation de gaz pauvres de gazogènes dans nos mo-*
    1975   *teurs à combustion interne type dual-fuel.*

63. Dvoskin, D. & Heady, E.O. *U.S. agricultural production under limited energy*
    1976   *supplies, high energy prices, and expanding agricultural exports.* Ames,
    Iowa, Iowa State University. 163 p.

64. EATON, WILLIAM, A. *Geothermal energy.* Washington, D.C. Energy Research
    1975 and Development Agency, Office of Public Affairs. 35 p.

65. EDELSON, EDWARD. MHD generators: super blowtorches deliver more power
    1974 with less fuel. *Popular Science,* 206(3): 86-88.

66. EDELSON, E. New ways to make solar cells trim costs of future sun power
    1976 for your home. *Popular Science,* 208(5): 74-77, 152, 156.

67. ELLIS, A.J. Geothermal systems and power development. *American Scientist,*
    1975 63: 510-520.

68. ENTREPRISES AGRICOLES. *Machines agricoles: la chase aux gaspillages d'énergie,*
    1975 p. 65-67.

69. ENERGY RESEARCH AND DEVELOPMENT AGENCY (ERDA). *A national plan*
    1976 *for energy research, development and demonstration: creating energy
    choices for the future.* Vol. 1. *The plan.* Washington, D.C. 122 p.

70. ERDA. *Report of the Proceedings of the Agriculture Processing Industry*
    1976 *Workshop on Energy Conservation.* Washington, D.C. 83 p.

71. ERDA. *Report of the Proceedings of the Energy Research and Development*
    1976 *Administration Workshop on Energy Conservation in Agricultural Pro-
    duction.* Washington, D.C. 101 p.

72. EUROPEAN COMMISSION ON AGRICULTURE. *The use of energy in European agri-*
    1976. *culture.* Twentieth session, 17-23 June 1976, Rome, Italy.

73. EXELL, R.H.B., KORNSAKOO, SOMMAI & WIJERATNA, D.G.D.C. *The design*
    1976 *and development of a solar-powered refrigerator.* Bangkok, Asian Institute
    of Technology. 74 p. Research Report No. 62.

74. FAIRBANK, W.C. *Fuel from feces?* ASAE Paper No. PC 74-03. 12 p.
    1974

75. FAO. *Production yearbook.* Rome.
    1973

76. FAO. *Organic materials as fertilizers.* Rome. Soils Bulletin No. 27.
    1974

77. FAO. *Shifting cultivation and soil conservation in Africa.* Rome. Soils Bulletin
    1974 No. 24.

78. FAO. *FAO preliminary report on the pesticide supply/demand situation. Ad*
    1975 *Hoc* Government Consultation on Pesticides in Agriculture and Public
    Health, Rome, 7-11 April 1975.

79. FAO. *Information system on pesticide supply and demand. Ad Hoc* Government
    1975 Consultation on Pesticides in Agriculture and Public Health, Rome,
    7-11 April 1975.

80. FAO. Energy for agriculture in the developing countries. *Monthly Bulletin*
    1976 *of Agricultural Economics and Statistics,* 25(2): 1-8.

81. FAO. *The state of food and agriculture, 1976.* Rome.
    1977

82. FAO/UNEP. *Development of a programme promoting the use of organic*
    1976 *materials as fertilizers.* Rome. AGL/MISC 1.

83. FARMER, R.C. *Towards a cost-effective engine.* Paper presented at the Gas
    1973 Turbine International.

84. FEDERAL ENERGY ADMINISTRATION. *Economic impact of shortages on the fer-*
    1975 *tilizer industry.* A report prepared by Arthur D. Little, Inc., Cambridge,
    Mass.

85. FEDERAL ENERGY ADMINISTRATION. *Energy consumption in the food system.*
    1975 A report prepared for the Industrial Technology Office, Federal Energy
    Administration, by Booz, Allen and Hamilton, Inc.

86. FEDERAL ENERGY ADMINISTRATION. *Energy and U.S. agriculture: 1974 data*
    1976 *base.* Vol. 1. Washington, D.C., Economic Research Service. 260 p.

87. FOOD TASK FORCE. *A hungry world. The challenge to agriculture.* Division
    1974 of Agricultural Science, University of California. 323 p.

88. FORBES, IAN. *The nuclear debate: a call to reason.* Boston, Mass.
    1974

89. FREE, J. Solar cells. *Popular Science*, December 1974, p. 52-55, 120, 121.
    1974

90. FRIED, E.R. & SCHULTZE, C.L. eds. *Higher oil prices and the world economy.*
    1975 Washington, D.C., Brookings Institution. 284 p.

91. GAC, A. Reflexion sur l'économie du froid. *Revue générale du froid. Journées*
    1976 *françaises du froid,* Octobre 1976. 19 p.

92. GARG, H.P. & KRISHNAN, A. *Solar drying of agricultural products.* I. *Drying*
    1974 *of chillies in a solar cabinet dryer.* Jodhpur, Central Arid Zone Research
    Institute.

93. GARG, H.P. *Solar energy utilisation research.* Jodhpur, Central Arid Zone
    1975 Research Institute. 48 p.

94. GARG, H.P. & THANVI, K.P. Year-round performance studies on a solar
    1975 cabinet dryer at Jodhpur. In *Proceedings of the Seventh Annual Meeting
    of All India Solar Energy Working Group and Conference on the Utilization
    of Solar Energy, Department of Mechanical Engineering and College of
    Agricultural Engineering, Punjab Agricultural University, Ludhiana, India,
    November 13-14, 1975,* p. 142-146.

95. GARG, H.P. & AMINULLAH. *Bibliography of Indian applied solar energy research*
    1976 *(1950-1975).* Jodhpur, Central Arid Zone Research Institute. 38 p.

96. GARG, H.P. Solar energy research at the Central Arid Zone Research Institute,
    1976 Jodhpur. Achievements and future plans. *Annals of Arid Zone*, 15(3):
    228-246.

97. GATES, D.M. The flow of energy in the biosphere. *Scientific American*, 225(3):
    1971 89-100.

98. GILES, G.W. The reorientation of agricultural mechanization for the develop-
1975 ing countries: policies and attitudes for action programme. *In* FAO.
*The effects of farm mechanization on production and employment. Report
of the Expert Panel.* Rome.

99. GIRARDIER, J.P. & CLEMENT, M.S. Design and performance of a simple solar
pump for lift-irrigation purposes. *Annals of Arid Zone.* Central Arid
Zone Research Institute, Jodhpur, India. 14 p.

100. GIRARDIER, J.P. & MASSEN, H. *Utilization of solar energy for lifting water.*
1970 Document, International Solar Energy Conference, Melbourne, Australia.

101. GLASSTONE, SAMUEL. *Controlled nuclear fusion.* Oak Ridge, Tenn., ERDA,
1974 Technical Information Center. 84 p.

102. GOLDENBERG, TED. Personal conversation regarding thesis for the degree
1976 of Ph.D.

103. GONY, J.N. *Déchets solides: production, traitement, récupération des ordures
1972 menagères en France.* BRGM.

104. GOTAAS, H.B. *Composting.* Geneva, World Health Organization.
1956

105. GREEN, M.B. & McCULLOCH, A. Energy considerations in the use of her-
1976 bicides. *J. Sci. Fd Agric.,* 27: 95-100.

106. GULVIN, H.E. *Farm engines and tractors.* New York, McGraw-Hill. 397 p.
1953

107. GUPTA, M.C. Solar space heating and cooling. In *Industrial Applications
1975 of Solar Energy, Seminar Proceedings, National Productivity Council,
Madras, India, June 4-6, 1975.*

108. HARDY, R.W.F. & HAVELKA, U.D. Nitrogen fixation research: a key to
1975 world food? *Science,* 188: 633-643.

109. HARDY, R.W.F. & HAVELKA, U.D. *In* Nutman, P. *Symbiotic nitrogen fixation
1975 in plants.* London, Cambridge University Press. International Biological
Programme Series, Vol. 7.

110. HARDY, R.W.F. *In* Newton, W.E. and Nyman, D.J. *Symposium on Dinitrogen
Fixation.* Pullman, Wash., Washington State University Press. (In
press)

111. HAYAMI, Y. & RUTTAN, V.W. *Agricultural development: an international per-
1971 spective,* p. 57. Baltimore, Md., Johns Hopkins Press.

112. HAYWOOD, R.W. *Analysis of engineering cycles.* Oxford, Pergamon Press.
1967

113. HEICHEL, G.H. Energy needs and food yields. *Technological Review,* 76(8):
1974 2-9.

114. HEITLAND, H.H. The Volkswagen alternative fuel programs. In *Capturing
1976 the Sun Through Bioconversion. Proceedings. A Conference on Capturing the Sun through Bioconversion, Washington, D.C.,* p. 403-416.

115. HENAHAN, JOHN F. Geothermal energy, the prospects get hotter. *Popular*
1974 *Science,* 206(11): 96-99, 142-143.

116. HLAVEK, R. & DUPUIS, J. *L'utilisation de l'énergie éolienne pour l'exhaure*
1971 *de l'eau au Mali. Bilan d'exploration des éoliennes de la région de Gao.*
Bulletin de liaison. Comité inter-africain.

117. HOLLANDER, J.M. & SIMMONS, M.K. eds. *Annual Review of Energy.* Vol. I.
1976 Palo Alto, Calif., Annual Reviews. 793 p.

118. HORSFIELD, B.C. & WILLIAMS, R.V. *Producer gas generation.* Document,
1976 Rural Electric Conference, University of California, January 1976.

119. HUBBERT, M.K. *Nuclear energy and the fossil fuels: drilling and production*
1956 *practice.* New York, American Petroleum Institute.

120. HUBBERT, M.K. Energy resources. *In* National Research Council Committee
1969 on Resources and Man. *Resources and man.* San Francisco, Calif.,
Freeman.

121. HUBBERT, M.K. Energy resources for power plant production. *In* Seale,
1973 Robert L. and Sierka, Raymond A. *Energy needs and the environment,*
p. 9-52. Tucson, Ariz., University of Arizona Press. 349 p.

122. HUBBERT, M.K. *U.S. energy resources: a review as of 1972.* Washington,
1974 D.C., Committee on Interior and Insular Affairs, U.S. Senate, 93rd
Congress, 2nd Session.

123. HUGHES, HELEN & PEARSON, S. *Principal issues facing the world fertilizer*
1974 *economy. Report of a Seminar on Fertilizers.* Princeton, N.Y., Agricultural Development Council and the IBRD. 11 p.

124. IDYLL, C.P. The harvest of seaweed. *Sea Frontiers,* 17: 342-348.
1971

125. INSTITUT NATIONAL DE LA RECHERCHE AGRONOMIQUE. *Aspects économiques*
1976 *de la récupération des pailles. Rapport de synthèse.* Grignon, Labo-
ratoire d'économie rurale.

126. INTERNATIONAL BANK FOR RECONSTRUCTION AND DEVELOPMENT. *World bank*
1976 *atlas.* Washington, D.C.

127. INTERNATIONAL BANK FOR RECONSTRUCTION AND DEVELOPMENT. *Rural elec-*
1975 *trification.* Washington, D.C. 80 p.

128. GERMAN FOUNDATION FOR INTERNATIONAL DEVELOPMENT. *Appropriate tech-*
1975 *nologies for semiarid areas: wind and solar energy for water supply.*
Seminar Center for Economic and Social Development. 334 p.

129. *International petroleum yearbook.*
1972

130. INTERNATIONAL RICE RESEARCH INSTITUTE. *Inventory of natural resources.*
1976 *Energy resources.* Manila, Agricultural Engineering Department.

131. ISHINO, IVAO. Personal communication.
1977

132. ISMAN, M. *Utilisation de l'énergie solaire dans les pays en développement par*
1976 *l'intermédiaire de la photosynthèse et de la fermentation méthanique.*

133. JENNINGS, B.H. & ROGERS, W.L. *Gas turbine analysis and practice.* New
   1953 York, McGraw-Hill.

134. JOHN DEERE AND CO. *Fundamentals of service. Engines.* Moline, Ill.
   1968

135. JOINT COMMITTEE ON ATOMIC ENERGY. *Understanding the national energy*
   1973 *dilemma.* Washington, D.C., Center for Strategic and International
         Studies, Georgetown University.

136. JONGMA, JULES. *Wood for energy in developing countries.* Rome, FAO. (In
         press)

137. KAPUR, J.C. Socio-economic considerations in the utilization of solar energy
   1961 in under-developed areas. *Proceedings of the United Nations Conference*
         on New Sources of Energy, Vol. 1, p. 58-66.

138. KIMURA, KEN-ICHI. *Present technologies of solar heating, cooling and hot water*
   1976 *supply in Japan.* Tokyo, Department of Architecture, Waseda Uni-
         versity. Reprinted from *Architectural Science Review,* Vol. 19, No. 2.
         Surrey Hills, Research Publications Ptg. Ltd. 4 p.

139. KOENIG, H.E. *Resource management in a changing environment.* East Lansing,
   1976 Mich., Michigan State University. DMRE-76-15.

140. KRISHNAN, A. & GARG, H.P. *Windpower for pumping water.* Jodhpur, Central
   1975 Arid Zone Research Institute. 4 p.

141. LAPERROUSAZ, PIERRE. *Alcool de paille: une idée en fermentation.* Cahiers
   1975 des Ingénieurs agronomes, No. 304.

142. LARSON, DENNIS L. *Solar energy for irrigation pumping.* Tucson, Ariz., Uni-
   1977 versity of Arizona. ASAE Paper No. 77-4021. 14 p.

143. LAUER, D.A. Limitations of animal waste replacement for inorganic fertilizers.
   1975 *In* Jewell, W.J. *Energy, agriculture and waste management. Proceedings*
         of 1975 Cornell Agricultural Waste Management Conference, p. 409-432.
         Ann Arbor, Mich., Ann Arbor Science Publishers.

144. LEACH, G. & SLESSER, M. *Energy equivalents of network inputs to food pro-*
   1973 *ducing processes.* Glasgow, University of Strathclyde.

145. LEACH, G. *Energy and food production.* Guildford, IPC Science and Tech-
   1976 nology Press. 137 p.

146. LE SAUVAGE. *Le méthane: à la recherche du troisième souffle,* p. 86-91.
   1976

147. LOF, GEORGE. Use of solar energy for heating purposes: solar cooking.
   1961 *Proceedings of the United Nations Conference on New Sources of Energy,*
         Vol. 5, p. 304-315.

148. LOVINS, A.B. *World energy strategies.* Cambridge, Mass., Ballinger. 131 p.
   1975

149. LUCAS, R.E., HOLTMAN, J.B. & CONNOR, L.J. *Soil carbodynamics and cropping*
         practices. Proceedings of the Energy and Agricultural Conference, St.
         Louis, Missouri, 1976. New York, Academic Press.

150. LYERLY, R.L. & MITCHELL, W. *Nuclear power plants.* Oak Ridge, Tenn.,
     1974 U.S. Atomic Energy Commission. 55 p.

151. MAKHIJANI, A. & POOLE, ALAN. *Energy and agriculture in the third world.*
     1975 Cambridge, Mass., Ballinger. 168 p.

152. MALAYSIAN CENTER FOR DEVELOPMENT STUDIES. *Report on the International*
     1975 *Workshop on Energy, Resources and the Environment, Pulau, Penang,*
          Malaysia. 186 p.

153. MARX, JEAN L. Nitrogen fixation. Research efforts intensify. *Science*, 185:
     1974 132-136.

154. McCARTY, P.L. Anaerobic waste treatment fundamentals. Part II. Envi-
     1964 ronmental requirements and control. *Public Works*, 45.

155. McDOWELL, R.E. *Report of National Dairy Research Institute.* UNDP/Unesco/
     1975 ICAR Project 73/020/13.

156. McHALE, JOHN. *World facts and trends.* New York, Collier Books.
     1972

157. MEADOWS, D.H. *The limits to growth.* New York, Universe Books.
     1972

158. MEINEL, A.B. & MEINEL, M.P. *Applied solar energy: an introduction.* Reading,
     1976 Mass., Addison-Wesley. 651 p.

159. MERRIAM, MARSHAL F. *Wind energy for human needs.* Berkeley, Calif.,
     1974 Laurence Berkeley Laboratory, University of California. 99 p. UCID-
          3724.

160. MERRIL, R. ET AL. *Energy primer.* Menlo Park, Calif., Portola Institute.
     1974 200 p.

161. MILLER, D.L. Ethyl alcohol. In *Capturing the sun through bioconversion.*
     1976 *Proceedings. A Conference on Capturing the Sun through Bioconversion,*
          p. 441-448. Washington, D.C.

162. MINISTÈRE DE LA COOPÉRATION. Energie solaire: possibilités d'utilisation dans
     1975 les pays du Sahel. Documents d'étude 18. Sous-direction de l'écono-
          mie et de la planification. 44 p.

163. JAPAN. MINISTRY OF INTERNATIONAL TRADE AND INDUSTRY. *Japan's new*
     1976 *energy policy.* BI-18. *Background information.* Tokyo, Agency of
          Natural Resources and Energy. 87 p.

164. MORSE, R.N. ET. AL. *Solar energy as a major primary energy source. Proceedings*
     1973 *of the Symposium on Realistic Prospects for Solar Power in Australia.*
          International Solar Energy Society.

165. MOUMOUNI, A. *Small- and medium-scale applications of solar energy and their*
     1973 *potential for the developing countries.* Paris, Unesco, Working Party
          on Solar Energy. 19 p.

166. MOUMOUNI, A. *L'énergie solaire: solution future aux problèmes énergétiques,*
     1976 p. 15-19. Onersol. Niger, Africa.

167. NALEWAGA, J.D. Energy requirements of various weed control practices.
     1974 *Proceedings of the North Central Weed Control Conference,* 29: 19-23.

168. NATIONAL ACADEMY OF SCIENCES. *Solar energy in developing countries.* Wash-
     1972 ington, D.C.

169. NATIONAL ACADEMY OF SCIENCES. *Agricultural production efficiency.* Wash-
     1975 ington, D.C. 199 p.

170. NATIONAL RESEARCH COUNCIL. *World food and nutrition study: the potential*
     1977 *contributions of research.* Washington, D.C. 192 p.

171. NATIONAL COMMITTEE ON SCIENCE AND TECHNOLOGY. *Report of the Fuel and*
     1974 *Power Sector.* *Technology Bhawan.* New Delhi. 851 p.

172. NATIONAL PRODUCTIVITY COUNCIL. *Solar energy application in India: status*
     1973 *and potential. Total energy and energy substitution. Seminar Proceedings.*
     New Delhi. 282 p.

173. NATIONAL SCIENCE FOUNDATION. *Solar energy as a national energy resource.*
     1972 NASA Solar Energy Panel. Department of Mechanical Engineering,
     University of Maryland. 85 p.

174. NELSON, L.F., BURROWS, W.C. & STICKLER, F.C. *Recognizing production,*
     1975 *energy-efficient agriculture in the complex U.S. food system.* ASAE
     Paper No. 75-5505. 20 p.

175 NEULING, S. Der Wärmeaufwand für den Betrieb von Biogasanlagen. *Agrar-*
     1955 *technik,* 5 (6).

176. *Newsweek.* How safe is nuclear power? *Newsweek,* 12 April: 70-75.
     1976

177. NIERAT, J.M. *De l'électricité produite à partir du bois. Courrier de l'exploi-*
     1975 *tant et du scieur.* Paris, Centre technique du bois. 3 p.

178. NORTH, W.J. Giant kelp: sequoias of the sea. *National Geographic,* 142:
     1972 250-269.

179. ODUM, E.P. *Fundamentals of ecology.* 2nd ed. Philadelphia, Penn., Saunders.
     1959

180. ODUM, H.T. & ODUM, E.P. *Energy basis for man and nature.* New York,
     1976 McGraw-Hill. 297 p.

181. PAGNI, LUCIEN. *L'extension du rôle énergétique du bois.* Le Courrier, No. 29.
     1975 January-February.

182. PARADY, W.H. & TURNER, J.H. *Electric energy.* Athens, Ga., American
     1976 Association for Vocational Instructional Materials in cooperation with
     Georgia Power Company. 48 p.

183. PASSAT, J. *Energies pour aujourd'hui et demain.* 96 p.
     1975

184. PASSMORE, R. & DURNIN, J.V.G.A. Human energy expenditure. *Physiological*
     1955 *Reviews,* 35: 801-840.

185. PATEL, J.J. *The Gobar gas plant: its development, present status and future.*
1976 New Delhi, National Committee on Science and Technology.

186. PATTERSON, WALTER C. *Nuclear power.* Middlesex, Penguin Books. 304 p.
1976

187. PEYTURAUX, R. *L'énergie solaire. Energies nouvelles.* Paris, Premier Mi-
1976 nistre, Service d'information et de diffusion. No. 110. 26 p.

188. PICARD, J. Les légumineuses comme source de protéines pour les ruminants
1975 - leui intérêt particulier au travers de la crise de l'énergie. *Fourrages,*
juin 1975.

189. PIMENTEL, D. ET AL. Food production and the energy crisis. *Science,* 182:
1973 443-450.

190. PIMENTEL, DAVID, LYNN, W.R., MACREYNOLDS, W.F., HEWES, M.T. & RUSK, S.
1974 *Workshop on research methodology for studies of energy, food, man and
environment. Phase I.* Ithaca, N.Y., Center for Environmental Quality
Management, Cornell University. 52 p.

191. *Planejamento, Desenvolvimento. Brasil.* No. 32. p. 30-35.
1976

192. POLLARD, W.G. The long-range prospects for solar energy. *American Scientist,*
1976 64: 424-429.

193. POST, R.F. & RIBE, F.L. Fusion reactors as future energy sources. *Science,*
1974 186 (4162): 397-407.

194. POWERS, PHILIP N. *Nuclear power: potential for the short term.* Paper presented
1975 at the Food and Energy Symposium of the Farm Electrification Council,
Washington, D.C. 18 p.

195. PRASAD, C.R. ET AL. Inde, le recyclage des déchets du village, une solution
1975 à la pénurie d'énergie et d'engrais. Actuel développement. *Economic
and Political Weekly, Bombay,* September-October 1975: 42-45.

196. PRIEST, JOSEPH. *Energy for a technical society.* Reading, Mass., Addison-
1975 Wesley. 358 p.

197. PRIMACK, JOEL, ed. Nuclear reactor safety: a special report. *Bulletin of
1975 Atomic Scientists,* September 1975: 15-51.

198. RAMAKUMAR, R. Harnessing wind power in developing countries. School
1975 of Electrical Engineering, Oklahoma State University. Article No.
759145. *IECEC '75 Record,* p. 966-973.

199. RAO, D.P. & RAO, K.S. *Solar water pump for lift irrigation.* Pilani, Rajasthan,
1976 Department of Chemical Engineering, Birla Institute of Technology
and Science.

200. REINHOLD, F. & NOACK, W. Laboratoriumsversuche über die gasgewinnung
1956 aus landwirtschaftlichen Stoffen. *Münchner Beiträge zur Abwasser-,
Fischerei- und Flussbiologie,* 3.

201. REVELLE, ROGER. Energy use in rural India. *Science,* 192: 969-975.
1976

202. RILEY, JOHN G. & SMITH, NORMAN. *Solar energy utilization by photosynthetic*
     1977 *production of solid fuel.* Orono, Maine, University of Maine. ASAE
     Paper No. 77-4018. 22 p.

203. ROGOWSKI, A.R. *Elements of internal-combustion engines.* New York,
     1953 McGraw-Hill. 234 p.

204. ROLLER, W.L. *Growing organic matter as a fuel raw material.* NASA report
     1975 CR-2608. 30 p.

205. ROSEGGER, S. Energetische Fragen bei der biologischen Gaserzeugung in der
     1955 Landwirtschaft. *Agrartechnik*, 5 (10).

206. SAKR, IBRAHIM A. *An elliptical paraboloid solar cooker.* Document, Inter-
     1973 national Congress on the Sun in the Service of Mankind, Paris.

207. SANKAR, T.L. Domestic adjustment and accommodations to higher raw ma-
     1976 terial and energy prices. In *Changing resource problems of the fourth
     world*, p. 51-89. Washington, D.C., Resources for the Future.

208. SATHIANATHAN, M.A. *Biogas achievements and challenges.* New Delhi, As-
     1975 sociation of Voluntary Agencies for Rural Development. 192 p.

209. SAUERLANDT, W. & GROETZNER, E. Eigenschaften und Wirkungen der bei
     1956 der biologischen Gasgewinnung aus Stallmist anfallenden organischen
     Dunger. *Münchner Beiträge zur Abwasser-, Fischerei- und Flussbiologie*, 3.

210. SCHIPPER, L. Raising the productivity of energy utilization. In Hollander,
     1976 J.M. and Simmons, M.K. *Annual Review of Energy, 1: 455-518.* Palo
     Alto, Calif., Annual Reviews, Inc. 193 p.

211. SCHUMACHER, E.F. *Small is Beautiful.* New York, Harper and Row.
     1973

212. SCHURR, SAM H. ET AL. *Energy, economic growth and the environment.* Bal-
     1972 timore, Md., Johns Hopkins University Press. 232 p.

213. SECTORAL REPORTS OF THE ENERGY COMMITTEE. *Energy for development.*
     1976 Presented at the International Conference on the Survival of Human-
     kind: The Philippine Experiment, Philippine International Convention
     Center, Manila, Philippines. September 6-10, 1976. 101 p.

214. SELLERS, W.D. *Physical climatology.* Chicago, Ill., Chicago University Press.
     1965

215. SHEARER, A.R. *Wairakei Power Station. New Zealand electricity.* Wellington,
     1974 Government Printer. 4 p.

216. SHEPPARD, M.L., CHADDOCK, J.B., COCKS, F.H. & HARMAN, C.M. *Introduc-
     1976 tion to energy technology.* Ann Arbor, Mich., Ann Arbor Science Pub-
     lishers. 300 p.

217. SHORT, TED, ROLLER, W.L. & BADGER, P.C. *A solar pond for heating green-
     1976 houses and rural residences.* Ohio State University. ASAE Paper No.
     76-4012. 11 p.

218. SHOVE, GENE C., ed. *Proceedings of the Solar Grain Drying Conference, Uni-
     1977 versity of Illinois, Urbana-Champaign, Illinois, January 11-12,1971.* 281 p.

219. SINGH, C.B., KAHLON, S.S., RANDHAWA, G.S. & PANESAR, B.S. *Annual report:*
1975 *energy requirements in Intensive Agricultural Production Programme.*
ICAR Co-ordinated Project. Ludhiana, Department of Farm Power
and Machinery, Punjab Agricultural University.

220. SINGH, R.B. *Technical report on biogas plant.* Ajitmal, Etawah, U.P., Gobar
1975 Gas Research Station.

221. SMITH, R.J. & FEHR, R.L. *Methane from agricultural residue.* Document,
1976 New England Conference on Energy in Agriculture, May 3-4, 1976.
10 p.

222. Société Technique de Géothermie, 152, Avenue des Champs Elysées, 75008
Paris, France.

223. SORENSON, H.A. *Gas turbines.* New York, Ronald Press.
1951

224. SPANO, L.A. Enzymatic hydrolysis of cellulose wastes to glucose. In *Captur-*
1976 *ing the Sun through Bioconversion. Proceedings: A Conference on Capturing the Sun through Bioconversion.* Washington, D.C. 862 p.

225. STAUSS, W. Der heutige Stand der Biogasgewinnung aus landwirtschaftlichen
1956 Stoffen. In *Münchner Beiträge zur Abwasser-, Fischerei- und Flussbiologie,* 3.

226. STEINHART, C.E. & STEINHART, J.S. *Energy: sources, use and role in human*
1974 *affairs.* North Scituate, Mass., Dubary Press. 362 p.

227. STEINHART, C.E. & STEINHART, J.S. Energy use in the U.S. food system.
1974 *Science,* 184(4134): 307-316.

228. STICKLER, F.C., BURROWS, W.C. & NELSON, L.F. *Energy from sun, to plant,*
1975 *to man.* Moline, Ill., Deere and Co.

229. STOUT, B.A. & DOWNING, C.M. *Selective employment of labor and machines*
1974 *for agricultural production.* East Lansing, Michigan State University.
23 p. Monograph No. 3.

230. STOUT, B.A. & DOWNING, C.M. Agricultural mechanization policy. *Inter-*
1976 *national Labour Review,* 113(2): 171-187.

231. STOUT, B.A. & DOWNING, C.M. Increasing the productivity of human, animal
1977 and engine power. In *Food enough or starvation for millions?* International Association of Agriculture and Economics.

232. STOUT, B.A. ET AL. *Energy use in agriculture: now and for the future.* Council
1977 for Agricultural Science and Technology Report Ro. 68.

233. SUGGS, C.W. & SPLINTER, W.E. Effect of environment on the allowable
1961 workload of man. *ASAE Transactions,* 4(1): 48-51.

234. SUGGS, C.W. & SPLINTER, W.E. Some physiological responses of man to
1961 workload and environment. *Journal of Applied Physiology,* 16(3): 413-420.

235. SUMMER, C.M. The conversion of energy. *Scientific American,* September 1971:
1971 149-160.

236. SUNSHINE PROJECT PROMOTION HEADQUARTERS. *Japan's Sunshine Project.*
1975 Tokyo, Agency of Industrial Science and Technology, Ministry of International Trade and Industry. 52 p.

237. SUNSHINE PROJECT PROMOTION HEADQUARTERS. *Solar Energy R & D Pro-*
1975 *gram. Sunshine Project in Japan.* Tokyo, Agency of Industrial Science and Technology, Ministry of International Trade and Industry. 74 p.

238. TABOR, H.Z. A solar cooker for developing countries. *Solar Energy,* 19(4):
1966 153-157.

239. TABOR, H.Z. Solar ponds. *Science Journal,* June 1966: 56-61.
1966

240. TABOR, H.Z. *Solar energy for developing regions.* Working Party on Solar
1973 Energy. Paris, Unesco. 20 p.

241. TENNESSEE VALLEY AUTHORITY. *Fertilizer trends.* Bulletin Y-77.
1973

242. THIRRING, H. *Energy for man from windmills to nuclear power.* New York,
1958 Harper and Row. 439 p.

243. THORNDIKE, E.H. *Energy and environment: a primer for scientists and engineers.*
1976 Reading, Mass., Addison-Wesley. 286 p.

244. TSCHIERSCHKE, H. Die Erzeugung von Biogas im landwirtschaftlichen Betrieb.
1961-62 *Archiv fur Landtechnik,* 3 (H.3).

245. UNGER, S.G. Energy utilization in the leading energy consuming food pro-
1975 cessing industries. *Food Technology,* December 1975:33-45.

246. *Unitar News.* Technology development.
1974

247. UNITED NATIONS. *Rural electrification in Asia and the Far East.* New York
1963 E/CN.11/640. ST/TAO/SER.C/63. 72 p.

248. UNITED NATIONS. *Small-scale power generation.* New York, Department of
1967 Economic and Social Affairs. 215 p.

249. UNITED NATIONS. *World energy supplies, 1970-73.* Statistical Papers, Series J.
1970-73 No. 18.

249a. UNITED NATIONS. *World energy supplies, 1971-75.* Statistical Papers, Series J.
1977 No. 20.

250. UNITED NATIONS. ECONOMIC AND SOCIAL COUNCIL. *Proceedings of the twelfth*
1972 *session of the Sub-Committee on Energy Resources and Electric Power, Economic and Social Commission for Asia and the Pacific.* Bangkok. Energy Resources Development Series No. 11. 316 p.

251. UNITED NATIONS. ECONOMIC AND SOCIAL COUNCIL. *Biogas technology and*
1975 *utilization.* Committee on Industry, Housing and Technology, Economic and Social Commission for Asia and the Pacific, New Delhi, India, 13-18 October 1975.

274 | References

252. UNITED NATIONS. ECONOMIC AND SOCIAL COUNCIL. *Electric energy in Africa:*
1976 *development and prospects.* Economic Commission for Africa, second
African Meeting on Energy, 1-12 March 1976.

253. UNITED NATIONS. ECONOMIC AND SOCIAL COUNCIL. *Proceedings of the second*
1976 *session of the Committee on Natural Resources, Economic and Social
Commission for Asia and the Pacific.* Bangkok. Energy Resources
Development Series No. 15. 111 p.

254. UNITED NATIONS. ECONOMIC AND SOCIAL COUNCIL. *Environmental impact*
1976 *of energy development and utilization in Africa.* Accra, Economic Commission for Africa. Second African Meeting on Energy, 1-12 March
1976. 27 p.

255. UNITED NATIONS. ECONOMIC AND SOCIAL COUNCIL. *Prospects for utilization*
1976 *of nuclear power in Africa.* Accra, Economic Commission for Africa.
Second African Meeting on Energy, 1-12 March 1976. 8 p.

256. UNITED NATIONS. ECONOMIC AND SOCIAL COUNCIL. *Report of the Expert*
1976 *Working Group on the Use of Solar and Wind Energy.* Bangkok, Economic and Social Commission for Asia and the Pacific, Committee on
Natural Resources, third session. 37 p.

257. UNITED NATIONS. *Statistical yearbook*, p. 157-365. New York.
1972

258. UNITED NATIONS. *Statistical yearbook.* New York.
1973

259. UNITED NATIONS. WORLD FOOD CONFERENCE. *The world food problem: pro-*
1974 *posals for national and international action.* E/CONF. 65 (14): 45. Rome.

260. U.S. ATOMIC ENERGY COMMISSION. *Light water breeder reactor.* Oak Ridge,
Tenn.

261. U.S. BUREAU OF MINES.

262. U.S. BUREAU OF THE CENSUS (1972).

263. U.S. DEPARTMENT OF AGRICULTURE. *The U.S. food and fiber sector: energy*
1974 *use and outlook.* Prepared by the Economic Research Service of the
U.S. Department of Agriculture for the Committee on Agriculture and
Forestry. U.S. Senate. Committee Printing. 111 p.

264. U.S. DEPARTMENT OF AGRICULTURE. *United States and world fertilizer outlook,*
1974 *1974 and 1980.* Washington, D.C., Economic Research Service. 66 p.
Agricultural Economics Report No. 257.

265. U.S. DEPARTMENT OF AGRICULTURE. *World fertilizer situation.* Washington,
1974 D.C. Economic Research Service. WAS-5.

266. U.S. DEPARTMENT OF INTERIOR. *Energy perspectives.* Washington, D.C. 207 p.
1975

267. *U.S. News and World Report.* The spread of atomic electricity. 30 Sep-
1974 tember 1974.

268. USMANI, I.H. *Review of the impact of production and use of energy on the* 1976 *environment and the role of UNEP.* New York, United Nations Environment Program. 88 p.

269. VAN GILST, W.J. *Farm equipment and fossil fuel consumption in agriculture.* 1974 Rome, FAO.

270. VILLARS, RALPH. *Energy and the fuel cell.*

271. VOLUNTEERS IN TECHNICAL ASSISTANCE. *Village technology handbook.* Mt. 1970 Rainier, Md.

272. VON OPPEN, M. *The sun basket.* Hyderbad, International Crops Research 1976 Institute for the Semi-Arid Tropics.

273. WARD, B. & DUBOIS, RENÉ. *Only one earth.* New York, Norton. 1972

274. WARD, G.T. *Energy as a major factor in man's development.* ASAE Paper 1970 No. 70-112. 22 p.

275. WATT, S.B. *A manual on the hydraulic ram for pumping water.* Silsoe, Bedford, 1975 Intermediate Technology Development Group. 38 p.

276. WEINBERG, ALVIN M. Global effects of man's production of energy. *Science,* 1974 186(4160): 205.

277. WEINGART, J.M. *Photovoltaic conversion of sunlight to electricity: considerations* 1975 *for developing countries.* International Institute for Applied Systems Analysis, Austria. 116 p.

278. WEISS, C. & PAK, SIMON. *Developing country applications of photovoltaic cells.* 1976 Paper presented at the National Solar Voltaic Program Review Meeting. 21 p.

279. WELLONS, L. *Using wood waste recovery methods in agriculture.* Paper, Rural 1976 Electric Conference. University of California.

280. WILCOX, H.A. *The ocean food and energy project.* A paper presented at the 1975 41st Annual Meeting of the American Association for the Advancement of Science.

281. WITTWER, SYLVAN H. *Biological nitrogen fixation.* Paper presented at the 1976 First Annual Review Meeting of the Fertilizer I.N.P.U.T.S. Project. Honolulu, East-West Food Review Institute of the East-West Center. Hawaii. 11 p.

282. ZARINCHANGE, J. *The Stirling engine.* London, Intermediate Technology 1972 Development Group. 32 p.

INDEX

Afghanistan

    per caput income, 33
    population, 33

Africa

    electrification, 167, 173–177
    energy consumption, 6, 7
    energy flow in agriculture, 44–45, 47–49, 52, 55, 57, 60
    energy reserves, 20, 21, 25
    fertilizer, 126, 132–133, 136–137, 139–142
    pesticides, 224
    straw production, 192
    waterpower, 19, 20
    wood fuel use, 190

Agricultural development, 39–45
    modern farming, 40–45
    shifting cultivation, 39
    traditional farming, 39–42, 44–45
    transitional farming, 40–45

Alcoholic fermentation, 204–205, 246

Algeria

    electrification, 167, 174, 175, 191
    fertilizer, 136
    per caput income, 32
    population, 32

Alternating-current generator, 159–160

Anaerobic fermentation, 199–204

Angola

    per caput income, 33
    population, 33

Animal draught power, 1, 3–5

Animal waste, 127–129, 138, 140, 142, 191, 193

Argentina

    energy consumption, 7, 8, 31
    gross national product, 31
    nuclear power, 179
    per caput income, 32
    population, 32
    wood fuel use, 19

Asia

    electrification, 167
    energy consumption, 3, 6
    energy reserves, 20, 21
    energy flow in agriculture, 42–44
    fertilizer, 126, 132, 133, 136, 137, 139, 140, 141, 142
    pesticides, 224
    straw production, 192
    waterpower, 19, 20
    wood fuel use, 190

Austria

    energy consumption, 35
    national income, 35

Australia

    energy consumption, 7, 31
    fertilizer, 136
    gross national product, 31
    waterpower, 20

Belgium

    energy consumption, 31

FAO SALES AGENTS AND BOOKSELLERS

Algeria
Société nationale d'édition et de diffusion, 92, rue Didouche Mourad, Algiers.

Argentina
Editorial Hemisferio Sur S.A., Librería Agropecuaria, Pasteur 743, 1028 Buenos Aires.

Australia
Hunter Publications, 58A Gipps Street, Collingwood, Vic. 3066; Australian Government Publishing Service, P.O. Box 84, Canberra, A.C.T. 2600; and Australian Government Service Bookshops at 12 Pirie Street, Adelaide, S.A.; 70 Alinga Street, Canberra, A.C.T.; 162 Macquarie Street, Hobart, Tas.; 347 Swanson Street, Melbourne, Vic.; 200 St. Georges Terrace, Perth, W.A.; 309 Pitt Street, Sydney, N.S.W.; 294 Adelaide Street, Brisbane, Qld.

Austria
Gerold & Co., Buchhandlung und Verlag, Graben 31, 1011 Vienna.

Bangladesh
ADAB, 79 Road 11A, P.O. Box 5045, Dhanmondi, Dacca.

Belgium
Service des publications de la FAO, M.J. de Lannoy, 202, avenue du Roi, 1060 Brussels. CCP 000-0808993-13.

Bolivia
Los Amigos del Libro, Perú 3712, Casilla 450, Cochabamba; Mercado 1315, La Paz; René Moreno 26, Santa Cruz; Junín esq. 6 de Octubre, Oruro.

Brazil
Livraria Mestre Jou, Rua Guaipá 518, São Paulo 05089; Rua Senador Dantas 19-S205/206, 20.031 Rio de Janeiro; PRODIL, Promoção e Dist. de Livros Ltda., Av. Venâncio Aires 196, Caixa Postal 4005, 90.000 Porto Alegre; A NOSSA LIVRARIA, CLS 104, Bloco C, Lojas 18/19, 70.000 Brasilia, D.F.

Brunei
SST Trading Sdn. Bhd., Bangunan Tekno No. 385, Jln 5/59, P.O. Box 227, Petaling Jaya, Selangor.

Canada
Renouf Publishing Co. Ltd, 2182 St Catherine West, Montreal, Que. H3H 1M7.

Chile
Tecnolibro S.A., Merced 753, entrepiso 15, Santiago.

China
China National Publications Import Corporation, P.O. Box 88, Beijing.

Colombia
Editorial Blume de Colombia Ltda., Calle 65 N° 16-65, Apartado Aéreo 51340, Bogotá D.E.

Costa Rica
Librería, Imprenta y Litografía Lehmann S.A., Apartado 10011, San José.

Cuba
Empresa de Comercio Exterior de Publicaciones, O'Reilly 407 Bajos entre Aguacate y Compostela, Havana.

Cyprus
MAM, P.O. Box 1722, Nicosia.

Czechoslovakia
ARTIA, Ve Smeckach 30, P.O. Box 790, 111 27 Prague 1.

Denmark
Munksgaard Export and Subscription Service, 35 Nørre Søgade, DK 1370 Copenhagen K.

Dominican Rep.
Fundación Dominicana de Desarrollo, Casa de las Gárgolas, Mercedes 4, Apartado 857, Zona Postal 1, Santo Domingo.

Ecuador
Su Librería Cía. Ltda., García Moreno 1172 y Mejía, Apartado 2556, Quito; Chimborazo 416, Apartado 3565, Guayaquil.

El Salvador
Librería Cultural Salvadoreña S.A. de C.V., Calle Arce 423, Apartado Postal 2296, San Salvador.

Finland
Akateeminen Kirjakauppa, 1 Keskuskatu, P.O. Box 128, 00101 Helsinki 10.

France
Editions A. Pedone, 13, rue Soufflot, 75005 Paris.

Germany, F.R.
Alexander Horn Internationale Buchhandlung, Spiegelgasse 9, Postfach 3340, 6200 Wiesbaden.

Ghana
Fides Enterprises, P.O. Box 14129, Accra; Ghana Publishing Corporation, P.O. Box 3632, Accra.

Greece
G.C. Eleftheroudakis S.A., International Bookstore, 4 Nikis Street, Athens (T-126); John Mihalopoulos & Son S.A., International Booksellers, 75 Hermou Street, P.O. Box 73, Thessaloniki.

Guatemala
Distribuciones Culturales y Técnicas "Artemis", 5a. Avenida 12-11, Zona 1, Apartado Postal 2923, Guatemala.

Guinea-Bissau
Conselho Nacional da Cultura, Avenida da Unidade Africana, C.P. 294, Bissau.

Guyana
Guyana National Trading Corporation Ltd, 45-47 Water Street, P.O. Box 308, Georgetown.

Haiti
Librairie "A la Caravelle", 26, rue Bonne Foi, B.P. 111, Port-au-Prince.

Hong Kong
Swindon Book Co., 13-15 Lock Road, Kowloon.

Hungary
Kultura, P.O. Box 149, 1389 Budapest 62.

Iceland
Snaebjörn Jónsson and Co. h.f., Hafnarstraeti 9, P.O. Box 1131, 101 Reykjavik.

India
Oxford Book and Stationery Co., Scindia House, New Delhi 110001; 17 Park Street, Calcutta 700016.

Indonesia
P.T. Sari Agung, 94 Kebon Sirih, P.O. Box 411, Djakarta.

Iraq
National House for Publishing, Distributing and Advertising, Jamhuria Street, Baghdad.

Ireland
The Controller, Stationery Office, Dublin 4.

Italy
Distribution and Sales Section, Food and Agriculture Organization of the United Nations, Via delle Terme di Caracalla, 00100 Rome; Libreria Scientifica Dott. Lucio de Biasio "Aeiou", Via Meravigli 16, 20123 Milan; Libreria Commissionaria Sansoni S.p.A. "Licosa", Via Lamarmora 45, C.P. 552, 50121 Florence.

Japan
Maruzen Company Ltd, P.O. Box 5050, Tokyo International 100-31.

Kenya
Text Book Centre Ltd, Kijabe Street, P.O. Box 47540, Nairobi.

Kuwait
Saeed & Samir Bookstore Co. Ltd, P.O. Box 5445, Kuwait.

Luxembourg
Service des publications de la FAO, M.J. de Lannoy, 202, avenue du Roi, 1060 Brussels (Belgium).

FAO SALES AGENTS AND BOOKSELLERS

Malaysia	SST Trading Sdn. Bhd., Bangunan Tekno No. 385, Jln 5/59, P.O. Box 227, Petaling Jaya, Selangor.
Mauritius	Nalanda Company Limited, 30 Bourbon-Street, Port Louis.
Mexico	Dilitsa S.A., Puebla 182-D, Apartado 24-448, Mexico 7, D.F.
Morocco	Librairie "Aux Belles Images", 281, avenue Mohammed V, Rabat.
Netherlands	Keesing Boeken V.B., Joan Muyskenweg 22, 1096 CJ Amsterdam.
New Zealand	Government Printing Office. Government Printing Office Bookshops: Retail Bookshop, 25 Rutland Street, Mail Orders, 85 Beach Road, Private Bag C.P.O., Auckland; Retail, Ward Street, Mail Orders, P.O. Box 857, Hamilton; Retail, Mulgrave Street (Head Office), Cubacade World Trade Centre, Mail Orders, Private Bag, Wellington; Retail, 159 Hereford Street, Mail Orders, Private Bag, Christchurch; Retail, Princes Street, Mail Orders, P.O. Box 1104, Dunedin.
Nigeria	University Bookshop (Nigeria) Limited, University of Ibadan, Ibadan.
Norway	Johan Grundt Tanum Bokhandel, Karl Johansgate 41-43, P.O. Box 1177 Sentrum, Oslo 1.
Pakistan	Mirza Book Agency, 65 Shahrah-e-Quaid-e-Azam, P.O. Box 729, Lahore 3.
Panama	Distribuidora Lewis S.A., Edificio Dorasol, Calle 25 y Avenida Balboa, Apartado 1634, Panama 1.
Paraguay	Agencia de Librerías Nizza S.A., Tacuarí 144. Asunción.
Peru	Librería Distribuidora "Santa Rosa", Jirón Apurímac 375, Casilla 4937, Lima 1.
Philippines	The Modern Book Company Inc., 922 Rizal Avenue, P.O. Box 632, Manila.
Poland	Ars Polona, Krakowskie Przedmiescie 7, 00-068 Warsaw.
Portugal	Livraria Bertrand, S.A.R.L., Rua João de Deus, Venda Nova, Apartado 37, 2701 Amadora Codex; Livraria Portugal, Dias y Andrade Ltda., Rua do Carmo 70-74, Apartado 2681, 1117 Lisbon Codex; Edições ITAU, Avda. da República 46/A-r/c Esqdo., Lisbon 1.
Korea, Rep. of	Eul-Yoo Publishing Co. Ltd, 46-1 Susong-Dong, Jongro-Gu, P.O. Box Kwang-Wha-Moon 362, Seoul.
Romania	Ilexim, Calea Grivitei N° 64-66, B.P. 2001, Bucharest.
Saudi Arabia	The Modern Commercial University, P.O. Box 394, Riyadh.
Sierra Leone	Provincial Enterprises, 26 Garrison Street, P.O. Box 1228, Freetown.
Singapore	MPH Distributors (S) Pte. Ltd, 71/77 Stamford Road, Singapore 6; Select Books Pte. Ltd, 215 Tanglin Shopping Centre, 19 Tanglin Road, Singapore 1024; SST Trading Sdn. Bhd., Bangunan Tekno No. 385, Jln 5/59, P.O. Box 227, Petaling Jaya, Selangor.
Somalia	"Samater's", P.O. Box 936, Mogadishu.
Spain	Mundi Prensa Libros S.A., Castelló 37, Madrid 1; Librería Agrícola, Fernando VI 2, Madrid 4.
Sri Lanka	M.D. Gunasena & Co. Ltd, 217 Olcott Mawatha, P.O. Box 246, Colombo 11.
Sudan	University Bookshop, University of Khartoum, P.O. Box 321, Khartoum.
Suriname	VACO n.v. in Suriname, Dominee Straat 26, P.O. Box 1841, Paramaribo.
Sweden	C.E. Fritzes Kungl. Hovbokhandel, Regeringsgatan 12, P.O. Box 16356, 103 27 Stockholm.
Switzerland	Librairie Payot S.A., Lausanne et Genève; Buchhandlung und Antiquariat Heinimann & Co., Kirchgasse 17, 8001 Zurich.
Thailand	Suksapan Panit, Mansion 9, Rajadamnern Avenue, Bangkok.
Togo	Librairie du Bon Pasteur, B.P. 1164, Lomé.
Tunisia	Société tunisienne de diffusion, 5, avenue de Carthage, Tunis.
United Kingdom	Her Majesty's Stationery Office, 49 High Holborn, London WC1V 6HB (callers only); P.O. Box 569, London SE1 9NH (trade and London area mail orders); 13a Castle Street, Edinburgh EH2 3AR; 41 The Hayes, Cardiff CF1 1JW; 80 Chichester Street, Belfast BT1 4JY; Brazennose Street, Manchester M60 8AS; 258 Broad Street, Birmingham B1 2HE; Southey House, Wine Street, Bristol BS1 2BQ.
Tanzania, United Rep. of	Dar es-Salaam Bookshop, P.O. Box 9030, Dar es-Salaam; Bookshop, University of Dar es-Salaam, P.O. Box 893, Morogoro.
United States of America	UNIPUB, 345 Park Avenue South, New York, N.Y. 10010.
Uruguay	Librería Agropecuaria S.R.L., Alzaibar 1328, Casilla de Correos 1755, Montevideo.
Venezuela	Blume Distribuidora S.A., Gran Avenida de Sabana Grande, Residencias Caroni, Local 5, Apartado 50.339, 1050-A Caracas.
Yugoslavia	Jugoslovenska Knjiga, Trg. Republike 5/8, P.O. Box 36, 11001 Belgrade; Cankarjeva Zalozba, P.O. Box 201-IV, 61001 Ljubljana; Prosveta, Terazije 16, P.O. Box 555, 11001 Belgrade.
Zambia	Kingstons (Zambia) Ltd, Kingstons Building, President Avenue, P.O. Box 139, Ndola.
Other countries	Requests from countries where sales agents have not yet been appointed may be sent to: Distribution and Sales Section, Food and Agriculture Organization of the United Nations, Via delle Terme di Caracalla, 00100 Rome, Italy.

Tipo-lito SAGRAF - Napoli